# Electrician's Handbook: The ideal companion for every technician.

"From Basic Principles to Civil Installations: a comprehensive course with small repairs and practical examples."

By Filippo Ferrari

Copyright © 2024 by Filippo Ferrari

All rights reserved.

No part of this book may be reproduced in any form without the written permission of the publisher or author, except as permitted by American copyright law

# Chapter 1: Fundamentals of Electricity ........... 6
## 1.1 Introduction to the concepts of voltage, current and resistance ........... 6
## 1.2 Ohm's and Kirchhoff's laws and their application . 7
## 1.3 Series and parallel circuits ........... 10
## 1.4 Measurement of electrical quantities ........... 14
## 1.5 Circuit analysis with practical examples ........... 18

# Chapter 2: Tools and Security ........... 24
## 2.1 Overview of basic power tools and techniques for proper use of tools ........... 24
## 2.2 Safety regulations in the working environment .... 39
## 2.3 Prevention of common electrical hazards ........... 40
## 2.4 Case studies on accidents and how to avoid them ........... 42

# Chapter 3: Design of Civil Installations ........... 45
## 3.1 Design principles of household electrical systems ........... 45
## 3.2 Plant planning and layout ........... 47
## 3.3 Choice of materials and components ........... 50
## 3.4 Domestic circuit design examples ........... 53

# Chapter 4: Installation of Installations ........... 59
## 4.1 Steps for safe installation of equipment ........... 59
## 4.2 Wiring and connection techniques ........... 61
## 4.3 Home automation and intelligent systems ........... 64
## 4.4 Verification and testing of installed systems ........... 67

### 4.5 Troubleshooting common problems during installation .................. 69

## Chapter 5: Maintenance and Control .................. 73

### 5.1 Troubleshooting common problems during installation .................. 73
### 5.2 Plant inspection techniques .................. 75
### 5.3 Technical documentation management .................. 78
### 5.4 Case studies of effective maintenance .................. 79

## Chapter 6: Fault Diagnosis .................. 83

### 6.1 Methodologies for fault identification .................. 83
### 6.2 Fault tracing and mapping techniques .................. 85
### 6.3 Repairing frequent faults .................. 88
### 6.4 Analysis of real cases and solutions adopted .................. 91

## Chapter 7: Common Repairs .................. 94

### 7.1 Guide to small home repairs .................. 94
### 7.2 Replacement of switches and sockets .................. 96
### 7.3 Repair of small household appliances .................. 99
### 7.4 Interventions on lighting and lamps .................. 102
### 7.5 Precautions and practical safety tips .................. 105

## Chapter 8: Practical Projects .................. 108

### 8.1 Basic circuit installation project .................. 108
### 8.2 Design and implementation of an alarm system 111
### 8.3 Garage or small office automation .................. 113
### 8.4 Electrical retrofit of a room .................. 116

## Chapter 9: Innovations and the Future of Electricity .. 120

9.1 Home Automation and Smart Home......................120

9.2 Renewable Energy ...........................................121

9.3 Electric Vehicle Charging Stations.........................122

9.4 Intelligent Lighting .........................................123

9.5 Home Automation ..........................................124

**Frequently Asked Questions in the Field of Electrician with Practical Examples**......................................126

Conclusion .........................................................130

# Chapter 1: Fundamentals of Electricity

## 1.1 Introduction to the concepts of voltage, current and resistance

In any electrical system, voltage, current and resistance are the three fundamental concepts for any technician to understand. These principles form the basis of the electrician's daily work, influencing every installation, diagnosis or repair.

**Voltage:** Voltage, or potential difference, is the force that forces electrons to move along a circuit. It is measured in volts (V) and can be compared to the pressure that pushes water through a pipe. Voltage can be direct current (DC) or alternating current (AC), and the difference between the two is essential to know. DC maintains a constant flow in a single direction, while AC reverses its direction at a defined frequency (50 or 60 Hz). In civil installations, AC voltage is generally used to power household appliances.

**Current:** Current is the flow of electrons through a conductor and is measured in amperes (A). There are two main types of current: direct current, which flows in a single direction, and alternating current, which changes direction periodically. Alternating current is more common in household electrical systems, while direct current is often used in electronics and batteries. The amount of current flowing through a circuit depends on both the applied voltage and the resistance of the circuit.

**Resistance:** Resistance measures the opposition to current flow in a circuit, expressed in ohms (Ω). Conductive materials such as copper and aluminum have low resistance and thus allow easy current flow. In contrast, insulating materials such as plastic and rubber offer high resistance, preventing the passage of electrons. Resistance in electrical circuits is crucial for the operation of components such as resistors, which can limit or split the current in a circuit, allowing safe and efficient use of electricity.

Understanding how voltage, current and resistance relate to each other is essential for any technician, as these concepts are closely related according to Ohm's and Kirchhoff's laws. In the next section, we will examine how these laws explain the behavior of complex circuits, providing a theoretical basis for designing and analyzing electrical systems with confidence.

Understanding these principles is critical to addressing the next steps of design, installation, and maintenance, making the training course consistent and progressive.

## 1.2 Ohm's and Kirchhoff's laws and their application

Ohm's and Kirchhoff's laws are fundamental pillars in electrical engineering and play a crucial role in understanding and designing electrical circuits. Although these concepts are often associated with calculations and formulas, it is possible to explain their importance and applications without going into mathematical detail.

**Ohm's Law:** Ohm's law is one of the first things electronics students learn. It establishes a direct relationship between voltage, current and resistance within an electrical circuit. According to this law, the current flowing through a conductor between two points is directly proportional to the voltage between those two points and inversely proportional to the resistance of the conductor. In more practical terms, if you increase the voltage across an electrical component while keeping its resistance constant, the current flowing through it will increase. Similarly, if the resistance of a component increases while the voltage remains constant, the current will decrease. This law is fundamental in the design and diagnosis of electrical circuits because it allows electricians and engineers to predict how changes in voltage or resistance affect the behavior of a circuit.

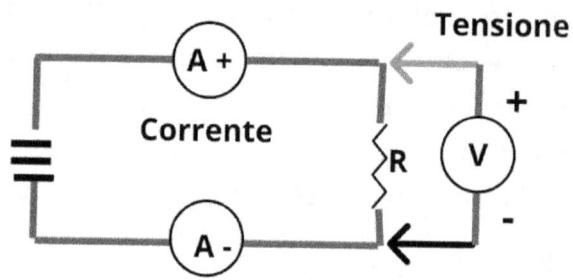

**V = Tensione**
**A = Corrente**
**R = Resistenza**

Kirchhoff's laws are essential for understanding and analyzing the behavior of electrical circuits. They are divided into two main categories, each of which helps us better understand how current and voltage are distributed in a circuit.

**Kirchhoff's First Law (Law of Nodes):** This law is based on the principle of conservation of electrical charge and states that the sum of currents flowing into a node is equal to the sum of currents flowing out of it. In other words, the amount of electricity entering an intersection point (node) in a circuit is exactly equal to the amount flowing out of it.

Formula della Prima Legge di Kirchhoff:

$$\sum I_{entranti} = \sum I_{uscenti}$$

Schema della Prima Legge di Kirchhoff:

In questo schema, le correnti $I_1$, $I_2$ e $I_3$ entrano nel nodo, mentre $I_4$ e $I_5$ escono dal nodo. Secondo la legge dei nodi, la somma di $I_1$, $I_2$, e $I_3$ sarà uguale alla somma di $I_4$ e $I_5$.

**Kirchhoff's Second Law (Mesh Law):** This law is based on the conservation of energy in a closed path or circuit mesh. It states that the algebraic sum of potential differences (voltages) encountered in a closed path is zero. This means that all applied voltages in a closed circuit are balanced by voltage drops within the same path.

Formula della Seconda Legge di Kirchhoff:

$$\sum V = 0$$

**Outline of Kirchhoff's Second Law:**

In the schematic, as we travel along the mesh in the direction shown, we encounter positive voltages generated by sources (e.g., batteries) and negative voltage drops across components such as resistors. The sum of these voltages along the closed path will be zero.

These two Kirchhoff's laws are fundamental to circuit analysis and enable engineers and technicians to predict circuit behavior and solve complex problems related to currents and voltages in larger electrical systems.

**Practical Applications:** In real life, Ohm's and Kirchhoff's laws find application in almost every aspect related to electrical circuits. From home systems to advanced electronic systems in industry, understanding how voltage, current and resistance interact with each other and how currents and voltages are distributed in a circuit is critical. Electricians use these laws to design safe and efficient circuits, to diagnose faults, and to optimize the performance of electrical systems.

In summary, Ohm's and Kirchhoff's laws not only help to understand and predict the behavior of electrical circuits but are also indispensable tools in the toolbox of every electrician and electrical engineer, essential for ensuring the safety and effectiveness of electrical systems.

In the next section, we will examine how series and parallel circuits operate according to these laws, highlighting the unique characteristics of each type of connection. Understanding these differences will help determine the best wiring scheme for each project, ensuring that systems are designed with efficiency, safety and reliability in mind.

Practical knowledge of these laws also becomes crucial for making accurate diagnoses and quick repairs, providing a solid framework for dealing with real circuits with confidence.

## 1.3 Series and parallel circuits

Understanding series and parallel circuits is critical for anyone working in electronics and electrical engineering, because these two types of configurations have significant impacts on the behavior of a circuit. Knowledge of how series and parallel circuits work helps in designing efficient and safe electrical systems, as well as diagnosing and solving problems.

**Series Circuits:** A series circuit is a configuration in which components are connected one after the other along a single path. In a series circuit, the same current flows through all components, from first to last. This means that if one component fails or is removed, breaking the circuit, current will stop flowing throughout the circuit, just as it would in a chain if one link broke. This aspect makes series circuits sensitive to interruptions.

A common example of a series circuit can be found in older Christmas lights, where if one bulb burns out, the whole string of lights stops working. Some types of old table lamps or chandeliers: some older light fixtures may have bulbs connected in series, especially those that use low-voltage bulbs.

Electronic toys and small devices: some toys and small electronic gadgets may have components (such as LEDs or small bulbs) connected in series to limit the current flowing through them. This happens because the circuit is broken and current can no longer flow through it.

**Parallel circuits:** Unlike series circuits, parallel circuits are configured with two or more paths for current flow. Components connected in parallel have separate connections to both the voltage source and return, allowing the current to divide its path between the different paths. This means that if a component in one parallel branch fails, it will not interrupt the flow of current in the other branches.

Parallel circuits are common in domestic and industrial applications, where safety and reliability are essential. For example, in a home, electrical outlets are connected in parallel so that the failure of one appliance does not affect the operation of the others.

loop

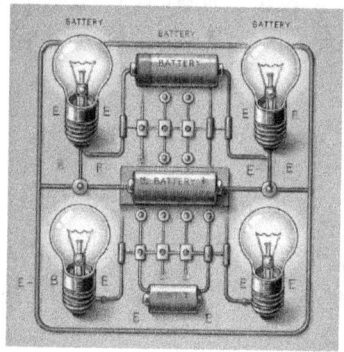
parallel circuit

Based on what criteria do you decide to apply a series or parallel circuit?

## 1. Independence of the Components

- **Parallel**: If one component fails, the others can continue to operate independently. This is advantageous in domestic and industrial environments where the failure of a single light point or device should not affect the rest of the system.

- **In series**: If one component fails, the entire circuit will stop working. This can be useful for safety applications where circuit failure can act as an alarm or interruption signal.

## 2. Voltage Distribution

- **Parallel**: Every component in the circuit receives the same voltage. This is ideal for household applications where devices such as lamps and sockets need a stable and uniform voltage.

- **In series**: The total circuit voltage is divided among the components. This can be useful in specific applications where it is necessary to reduce the voltage across individual components for their protection or optimal operation.

## 3. Security

- **Parallel**: Generally safer because a fault in a single branch does not interrupt the entire system. Also, short circuits in one branch do not affect other branches.

- **Bulk**: Less secure in terms of reliability because any interruption in one component can lead to a complete system failure.

## 4. Current Efficiency

- **Parallel**: Current can vary between branches depending on the load, allowing more efficient use of energy and better handling of heavy loads.

- **In series**: The current is the same in all components, which can limit functionality in scenarios where different devices require different current levels.

## 5. Ease of Maintenance

- **Parallel**: Easier to maintain and repair, since components can be added or removed without affecting others. This is especially useful in complex systems where frequent maintenance is necessary.

- **In series**: More difficult to maintain, as the addition or removal of one component affects the whole circuit.

### 6. Cost and Energy Efficiency

- **Parallel**: While it may require more wiring and protection (such as fuses or circuit breakers for each branch), it offers greater energy efficiency and better load balancing.
- **Bulk**: Less expensive in terms of wiring, but less energy efficient and with higher risks of overloading and component damage.

### Conclusion

The choice between series and parallel circuits depends on the specific application, safety and maintenance needs, and energy efficiency considerations. In modern residential and commercial electrical systems, parallel circuits are often preferred because of their flexibility and safety. However, knowing both types of circuits is essential for any electrical professional to apply the most appropriate solution for each specific situation.

In the next few points, we will explore how to effectively measure electrical quantities, an essential skill for verifying the proper functioning of circuits and diagnosing any problems.

### 1.4 Measurement of electrical quantities

Accurately understanding and measuring voltage, current, resistance, and other electrical variables is critical to the design, maintenance, and diagnosis of problems in electrical systems.

1. **Importance of measurement:** Measuring electrical quantities allows specialists to verify whether an electrical system is properly installed, operating efficiently and safely, and complying with applicable regulations. Measurements are also useful for locating faults, evaluating the performance of electrical devices and components, and performing preventive maintenance.

2. **Measurement tools:**

    - **Multimeter:** A versatile tool that can measure voltage, current and resistance. It is one of the most common and indispensable tools for electricians, available in both analog and digital versions.

    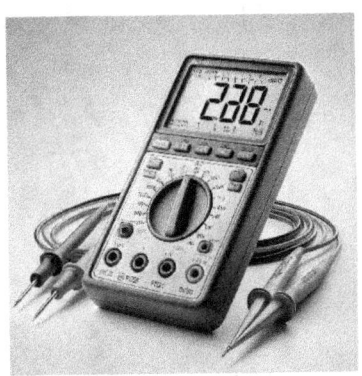

    - **Current Clamp:** Useful for measuring current in a circuit without the need to interrupt current flow. The current clamp is ideal for checking running systems because it does not require disconnecting cables.

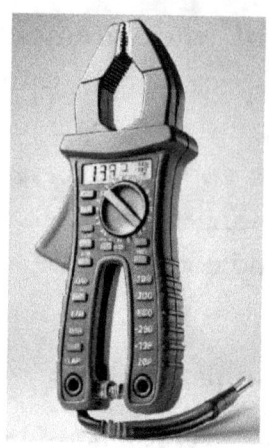

- **Voltage tester:** An instrument used to detect the presence of voltage in circuits. Voltage testers are critical for safety, allowing electricians to check whether a circuit is energized before working on it.

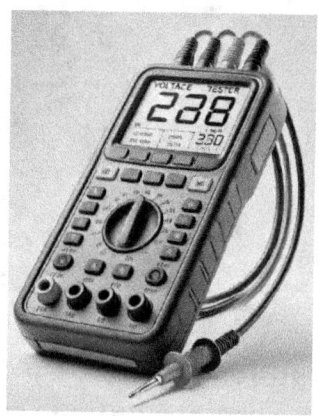

- **Oscilloscope:** This instrument is more sophisticated and is used to observe in detail the behavior of electrical waves in circuits. It is particularly useful in complex diagnosis and analysis of electronic systems.

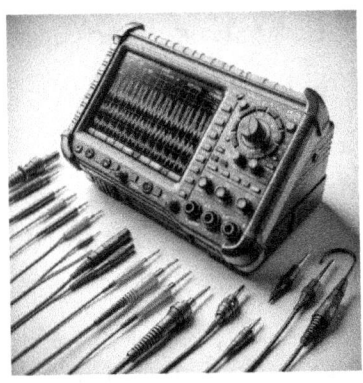

**3. Measurement procedure:** Making accurate measurements requires not only proper instruments, but also precise methodology. Here are some basic steps:

- **Preparation:** Make sure the measuring instruments are calibrated and in good working order. Check the battery and settings before starting measurements.

- **Safety:** Before measuring, it is essential to ensure that all safety rules are followed, such as wearing insulating gloves and making sure the work area is safe.

- **Function selection:** Choose the correct setting on the measuring device, depending on the quantity to be measured (voltage, current, resistance).

- **Instrument connection:** Place the probes or instrument terminals at the correct points in the circuit. For voltage measurements, connect to the two points between which you want to measure the potential difference; for current, insert the instrument in series with the circuit if you are not using a current clamp.

**4. Interpretation of results:** After taking measurements, it is important to interpret the results in the context of the electrical system under consideration. This may include comparison with expected values, analysis of possible causes of abnormal readings, and planning for possible corrective action.

**Conclusion:** The ability to accurately measure electrical quantities is a fundamental skill for electricians, necessary to ensure the safe and effective operation of electrical systems. As technology advances, measuring instruments are also evolving, offering increasing accuracy and new features to facilitate the work of specialists in the field.

## 1.5 Circuit analysis with practical examples

Circuit analysis is a crucial aspect of an electrician's or electrical engineer's job, enabling them to understand how the various components within a system work and how the different parts interact with each other. This process can be facilitated through the use of practical examples that illustrate the dynamics of circuits in real-world situations, without necessarily resorting to complex calculations or mathematical formulas.

**Basic concepts of circuit analysis:** Circuit analysis focuses on understanding the current flow and voltage distribution within an electrical system. It makes it possible to verify circuit functionality, identify possible problems, and optimize the configuration for greater efficiency and safety. This analysis is based on a few fundamental concepts:

- **Current flow:** Refers to the movement of electrons through a conductor. Analysis of current flow helps determine whether circuit components are receiving the amount of energy they need to function properly.

- **Voltage distribution:** Observing how the voltage is distributed across the various components of the circuit is critical to ensure that each part is operating within safe limits.

- **Integrity of components:** Check that all circuit components are in good condition and properly connected to prevent malfunctions or hazards.

**Practical examples of circuit analysis:**

Installing a new electric furnace requires a number of specific steps to ensure that it is safe, efficient, and compliant with electrical regulations. Here is a summary of the main steps and how to determine the proper capacity of the cord:

### 1) The Installation of an Electric Oven:

1. **Check Power and Circuit:** Check the oven specifications to make sure the existing circuit in your kitchen can support the load, or if a new dedicated circuit needs to be installed.
2. **Circuit Installation:** If necessary, install a new dedicated circuit from the electrical panel to the furnace location with a suitable circuit breaker.
3. **Preparing the Power Socket:** Install a suitable outlet near the oven and connect the wires according to regulations.
4. **Connecting the Oven:** Make sure the power is off before plugging the oven into the outlet. Follow the manufacturer's instructions for connecting the wires.
5. **Test the Oven:** Turn the power back on and test the oven to make sure it is working properly.

6. **Final Inspection**: Check that all connections are secure and monitor the kiln for signs of malfunction in the first few hours of use.

## Determine the Adequate Capacity of the Cable:

To ensure that the capacity of the cable is adequate to carry the current required by the furnace, it is important to consider two main factors:

- Cable Cross-Section: The cable cross-section must be sufficient to handle the maximum current that the furnace may require. This is based on the formula I=P/V (where I is current, P is power, and V is voltage). A cable with a larger cross section can handle more current without overheating.

- **Cable Type and Insulation**: The type of insulation and construction of the cable affect its ability to handle certain currents. Cables with higher quality insulation can better handle the heat generated by current flow.

- **Thermal Resistance**: Check that the thermal resistance of the cable is compatible with the installation environment. Cables exposed to high temperatures may need a larger cross section to ensure safety and durability.

Using cable load charts, often provided by manufacturers or included in local regulations, can help select the most suitable cable based on current, voltage and cable length.

## Image of the Connection Diagram:

In this diagram, you can see how the cable from the electrical panel (A) through the circuit breaker (B) carries power to the power outlet (D) and finally to the furnace (E), ensuring that everything is connected safely and in accordance with regulations.

With these steps and considerations, you can install your electric oven safely and efficiently, providing years of safe and reliable cooking.

## 2) Repair a washing machine that shows signs of burning or damage:

If your washing machine shows signs of damage or burning, it is important to proceed carefully and systematically to diagnose and fix the problem safely. Here are the steps to follow:

### 1. Disconnect the power supply

First, be sure to unplug the washing machine from the power outlet. This step is crucial to ensure your safety while inspecting and working on the appliance.

### 2. Identification of damaged areas

- **External visual examination**: Begin with a detailed visual examination of the washer's exterior, looking for signs of burning, discoloration, or deformation of the material, particularly around the control panel area and power outlets.

- **Interior inspection**: Remove the rear or bottom access panel (depending on model) to gain access to the inside of the washer. Look for signs of damage, such as burned wiring, black or misshapen components, and unusual burning smells.

### 3. Electronic control panel control

- **Check the control panel**: This component controls all functions of the washing machine and is often a source of failure if visually damaged. Look for signs of burn marks or damaged components on the circuit board.

- **Connections and wiring**: Check all electrical connections related to the control panel to make sure they are firm and undamaged. Burned or loose wiring should be replaced.

## 4. Inspection of the electronic board

- **Electronic** board: The electronic board may show signs of heat or overload damage. If you find burned components or broken circuits, the board will probably need to be replaced.

- **Board testing**: If you have a multimeter, you can test some components of the electronic board to check their functionality. However, this procedure may require specific technical knowledge.

## 5. Motor control

- **Motor inspection**: Also check the motor for any signs of overload or damage. Burned wiring and connectors to the engine may indicate more serious problems.

- **Motor operation**: If the motor shows signs of external damage, further diagnosis or replacement may be necessary.

## 6. Reassembly and testing

After replacing damaged parts and ensuring that everything is in order, carefully reassemble the washing machine. Make sure all components are in place and that the wiring is neat and secure.

## 7. Proof of function

- **Reintroducing power**: Reconnect the washing machine to the power supply.
- **Test**: Start a short wash cycle to check the operation of the machine. Watch the washing machine carefully during the cycle to make sure it is working properly and that there are no further signs of problems.

# Chapter 2: Tools and Security

## 2.1 Overview of basic power tools and techniques for proper use of tools

The multimeter is a versatile instrument used to measure various electrical quantities such as voltage, current and resistance. It can be extremely useful in many contexts, from basic electronics to solving complex problems in electrical systems. Here is how to use it through small steps:

### Step 1: Preparation

1. **Select the type of multimeter**: Make sure you have a multimeter suitable for the specific measurement you intend to make. Multimeters can be analog or digital.

2. **Set the multimeter**: Set the multimeter to the type of measurement you wish to make (volts, ohms, amps). Make sure the scale is appropriate for the magnitude you plan to measure to avoid damage to the meter.

### Step 2: Voltage Measurement

1. **Set the multimeter to volts**: Choose the setting V~ for alternating (AC) voltage or V- for direct (DC) voltage.

2. **Insert terminals**: Connect the black terminal to the common or ground and the red terminal to where you want to measure voltage.

3. **Measurement**: Touch the test points with the terminals and read the value on the display. Be sure not to touch the metal components of the terminals while measuring to avoid short circuits.

## Step 3: Current Measurement

1. **Set the multimeter to amperes**: Choose the setting A~ for alternating current or A- for direct current.

2. **Break the circuit**: To measure current, you must insert the multimeter in series in the circuit. This may require you to disconnect a circuit wire and connect the terminals of the multimeter to complete the run.

3. **Connect the terminals**: Make sure the connection is secure and the multimeter completes the circuit.

4. **Measure and reconnect**: Turn on the circuit and read the value on the multimeter. Remember to reconnect the circuit as it was originally after the measurement.

## Step 4: Measurement of Resistance

1. **Set the multimeter to ohms**: Turn the dial to the resistance position, often indicated by the symbol $\Omega$.

2. **Disconnect the power supply to the circuit**: Make sure the component or circuit has no power supply before measuring resistance.

3. **Connect the terminals**: Touch the test points with the terminals of the multimeter.

4. **Read the value**: Check the resistance on the display and make sure the circuit is disconnected from the power supply during the measurement.

## Step 5: Safety and Precautions

1. **Always check the setting**: Before each measurement, check that the multimeter is set correctly to avoid damage to the instrument or safety hazards.

2. **Use protection**: Always wear insulating gloves and protect your eyes when working with live circuits.

3. **Avoid direct contact**: Never touch exposed terminal parts or electrical components while the circuit is energized.

Using a multimeter takes practice and care, but by following these steps and always maintaining a cautious approach, you can get accurate measurements and work safely on electrical circuits.

A continuity tester, used to check whether a circuit is complete (i.e., whether there is a continuous path for electric current), can also be performed with the multimeter.

### Step 1: Preparation

1. **Select continuity tester**: Make sure you have a working continuity tester. Some multimeters also have a built-in continuity function.

2. **Check the battery**: Before starting, check that the continuity tester is properly powered and working properly. Most testers make a sound (beep) when they detect continuity.

### Step 2: Setting up the Tester

1. **Set the tester**: If you are using a multimeter, set the selector switch to the continuity test mode, often indicated by a continuity symbol (usually some sort of wave or sound symbol).

2. **Prepare the test leads**: Make sure the test leads are properly attached to the tester.

## Step 3: Continuity Test

1. **Disconnect the power supply**: Make sure the power supply to the circuit or device you are testing is disconnected to avoid damage to the tester or yourself.

2. **Touch the test points**: Apply the tester's probes to the two points on the circuit between which you wish to check continuity. It does not matter which probe touches which point, since continuity has no polarity.

3. **Listen for the beep**: A continuous sound from the tester indicates that there is continuity, that is, the circuit is closed and current can flow freely. If you hear no sound, it means the circuit is open (interruption).

## Step 4: Interpretation of Results

1. **Evaluation**: If the tester beeps, check that the connection is not too resistive, which could be a sign of a bad connection. Many continuity testers have a resistance threshold above which they will not sound.

2. **Troubleshoot**: If there is no continuity, visually check the circuit or component for breaks, broken wires, or cold solder.

## Step 5: Safety Precautions

1. **Handle with care**: Do not touch other conductive parts while testing continuity to avoid accidental short circuits.

2. **Insulation**: Use tools with insulated handles and always wear personal protective equipment when working with electrical circuits.

## Step 6: Maintenance of the Tester

1. **Cleaning**: After use, clean the probes and tester body to maintain accuracy and functionality.

2. **Storage**: Store the tester in a dry and safe place to avoid damage.

Using an uninterruptible power tester is a common practice for electricians during installation and maintenance of electrical systems, allowing quick verification of the integrity of connections. By following these simple steps, you can effectively use the continuity tester to ensure that your circuits are safe and functioning properly.

A current clamp is a very useful tool for measuring the current flowing in a circuit without the need to interrupt the current flow or make direct contact with conductors. It is ideal for checking currents in circuits that cannot be easily shut down or in situations that require a quick and safe measurement. Here's how to use it step by step:

## Step 1: Preparation

1. **Select the right current clamp**: Make sure the current clamp is appropriate for the type of current you intend to measure. Check the specifications to ensure that it covers both alternating current (AC) and direct current (DC), if necessary.

2. **Check functionality**: Before using it, check that the current clamp is in good condition and working properly.

## Step 2: Setting up the current clamp.

1. **Turn on the current clamp**: Activate the instrument using the power switch.

2. **Select the type of measurement**: Choose the correct function on the clamp, depending on whether you need to measure AC or DC current. Many current clamps have a selector switch to change between the two modes.

**Step 3: Current measurement**

1. **Open the clamp jaw**: Press the button or lever to open the clamp jaw.

2. **Place the jaw around the conductor**: Insert the cable inside the jaw of the pliers, being careful not to touch the cable with your hands. Make sure the jaw is fully closed around the conductor.

3. **Avoid measuring multiple cables at the same time**: To get an accurate reading, make sure the jaw of the clamp surrounds only a single conductor. The presence of multiple leads within the jaw can produce inaccurate readings.

**Step 4: Data reading and interpretation**

1. **Read the display**: After closing the jaw around the conductor, the current clamp will display the current flowing through it.

2. **It detects variations**: To check the stability of the current, observe the display for a few seconds. If the current varies significantly, you may need to investigate further to identify possible problems in the circuit.

**Step 5: Safety and precautions**

1. **Do not exceed instrument specifications**: Do not attempt to measure currents that exceed the maximum capacity of the current clamp, as this may damage the instrument or cause danger.

2. **Use protection**: Although the current clamp is a safe method of measurement because it does not require direct contact, it is always a good idea to wear insulating gloves and protect your eyes when working with electrical circuits.

## Step 6: Shutdown and maintenance

1. **Turn off the current clamp**: After completing the measurement, turn off the clamp to preserve battery life.

2. **Proper** storage: Store the current clamp in a protective case to avoid mechanical damage and maintain measurement accuracy over time.

Using a current clamp is an effective and safe way to measure current in electrical circuits, facilitating rapid diagnosis and risk-free maintenance.

The oscilloscope is a fundamental instrument for analyzing electrical signals that displays the waveform of time-varying signals. It is widely used in electronics, engineering and research laboratories. Here is how to use an oscilloscope through simple steps:

## Step 1: Preparation

1. **Select the right oscilloscope**: Make sure you have an oscilloscope that meets the needs of your specific

project, considering the bandwidth and sampling rate needed.

2. **Set up your workbench**: Place the oscilloscope in a stable location close to the device or circuit you intend to test.

## Step 2: Connection and Configuration

1. **Connect the device**: Connect the oscilloscope probe to the point on the circuit where you want to measure the signal. If you are measuring a high-frequency or very sensitive signal, use a compensated probe to avoid distortion.

2. **Ground the probe**: Connect the probe ground clip to a ground point in the circuit to avoid electrical interference.

3. **Turn on the oscilloscope**: Make sure the oscilloscope is turned off when you connect the probe. Turn it on only after you have made the connections.

## Step 3: Adjusting Basic Settings

1. **Select channel**: If your oscilloscope has multiple channels, choose the one you are using (often referred to as CH1, CH2, etc.).

2. **Set the time scale**: Adjust the time base (time/div) to display an appropriate time interval on the screen so that the signal can be clearly observed.

3. **Adjust voltage scale**: Adjusts the vertical scale (volts/div) to properly display the amplitude of the signal.

## Step 4: Visualization and Signal Analysis

1. **View the signal**: Observe the waveform on the display. Use the horizontal and vertical position controls to center the waveform on the screen.

2. **Use advanced functions**: If needed, use functions such as the cursor for precise voltage or time measurements, or the trigger function to stabilize periodic waveforms.

3. **Analyze waveform**: Identify signal characteristics such as frequency, amplitude and waveform. Compare your results with theoretically expected results to verify that the circuit is working properly or to diagnose problems.

## Step 5: Safety Precautions

1. **Handle with care**: Oscilloscopes are precise and sensitive instruments. Handle them with care and be sure not to overload the input.

2. **Proper insulation**: Make sure all connections are insulated and secure, especially when working with high voltages or high-energy circuits.

## Step 6: Maintenance and Shutdown

1. **Turn off the oscilloscope**: After use, turn off the oscilloscope and disconnect the probe.

2. **Cleaning and storage**: Clean the device and store it in a dry and safe place to preserve its longevity and performance.

By following these steps, you will be able to effectively use an oscilloscope to analyze electrical signals, enabling you to visualize and solve complex problems in a more informed and accurate way.

An impedance tester is an instrument used to measure impedance, which is the total resistance of a circuit to alternating current (AC) flow. Impedance combines resistance (resistive reactance) and reactance (capacitive and inductive reactance). This tool is very useful in analyzing audio circuits, testing power lines, and communication systems. Here is how to use an impedance tester step by step:

### Step 1: Preparation

1. **Select the appropriate impedance tester**: Make sure the tester is suitable for the specific impedance range you intend to measure and that it is compatible with the frequency of the current you are analyzing.

2. **Verify calibration**: Check that the tester is calibrated correctly to ensure the accuracy of measurements.

### Step 2: Connecting the Tester

1. **Turn off the power to the circuit**: Before connecting the tester, make sure the circuit or device you are testing is turned off to avoid damage to the instrument or personal injury.

2. **Connect the tester leads**: Connect the impedance tester leads to the points on the circuit where you wish to make the measurement. This may include specific terminals, leads, or components. If the tester has clips, make sure they are firmly attached for a secure connection.

### Step 3: Setting up the Tester

1. **Choose the type of measurement**: If the tester offers the ability to measure different types of impedances (such as series or parallel), select the appropriate setting based on your specific need.

2. **Set the frequency**: If the tester allows frequency adjustments, set it according to the frequency of the signal of the circuit you are testing.

## Step 4: Execution of the Measure

1. **Activate the tester**: Turn on the tester and start the measurement. The device will calculate the impedance and show the result on the display.

2. **Read the value**: Take note of the impedance value shown. Depending on the type of tester, you might also get information on specific resistance, capacitance, or inductance.

## Step 5: Analysis of Results

1. **Interprets the data**: Analyze the measured impedance value in relation to the specifications of the circuit or component being tested. Unexpected values may indicate problems such as short circuits, open circuits, or faulty components.

2. **Check consistency**: If necessary, repeat the measurement to confirm the consistency of the results, especially if the first results were unexpected or if you are diagnosing intermittent problems.

## Step 6: Safety Precautions

1. **Handle with care**: Electrical measuring instruments must be used with care to avoid damage to the instrument and risk to the operator.
2. **Insulation**: Make sure that all connections are secure and that the tester cables are in good condition to prevent the risk of electrical shock.

**Step 7: Maintenance and Storage**

1. **Turn off and disconnect**: After use, turn off the tester and disconnect the cables.
2. **Cleaning and storage**: Clean the tester if necessary and store it in a dry and safe environment to preserve its functionality and prolong its life.

Using an impedance tester requires an understanding of measurement principles and attention to detail. By following these steps, you will be able to accurately measure impedance and contribute to the effective analysis and maintenance of electrical circuits and devices.

In the field of electricity and electronics, wiring requires a number of specific tools to ensure safe and efficient installations. Here are some of the most common tools used for wiring:

1. **Crimping pliers**: Used to attach terminals to cables.

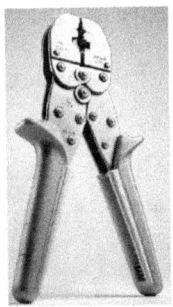

2. **Cutters**: These are used to cut cables.

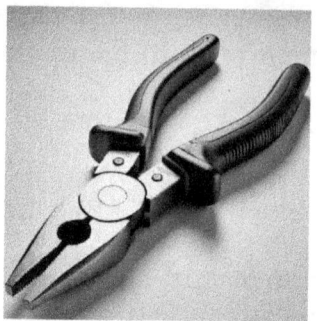

3. **Wire strippers**: Used to remove insulation from cables without damaging them.

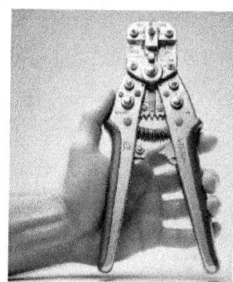

4. **Screwdrivers**: Needed to tighten screws in terminals or junction boxes.

5. **Voltage testers**: Used to check for voltage before starting work.

6. **Insulating tape**: Used to insulate cable connections or terminations.

7. **Cable ties**: Used to hold cables together neatly.

Among these, one of the most widely used tools is the **wire stripper**. This tool is essential for preparing wires for connection, ensuring that insulation is removed cleanly and precisely to avoid short circuits or other electrical problems.

# Operation of the Wire Stripper: Small Steps and Practical Examples

## Step 1: Preparation

- **Wire Stripper Selection**: Choose a wire stripper suitable for the type and size of cable you are working with. Wire strippers can be adjustable or have predefined recesses for different cable sizes.

## Step 2: Adjustment

- **Setting the Wire Stripper**: If the wire stripper is adjustable, adjust the opening of the blades according to the diameter of the wire to avoid cutting the inner wires.

## Step 3: Stripping the Cable

- **Placing the** Cable: Insert the cable into the recess corresponding to its size. Make sure it is firmly in place.
- **Stripping**: Close the handles of the wire stripper to cut the insulation. Then pull back the wire stripper to remove the insulation from the wire. This leaves the conductive wires exposed and ready to be connected or crimped.

## Practical Example

- **Connecting a Receptacle**: Suppose we need to connect a cable to an electrical outlet.
    - Measure the length of insulation to be removed, usually around 1 cm from the end of the cable.

- Use the wire stripper to remove the insulation without cutting or damaging the inner copper wires.

- Once exposed, the copper wires can be connected to the socket terminals. Make sure the connection is firm and secure.

## Step 4: Inspection

- **Checking the Cable:** After removing the insulation, inspect the cable to make sure there are no damaged wires. Broken copper filaments can affect the quality of the connection.

## Step 5: Cleaning and Maintenance

- **Care of the Wire Stripper:** Clean the wire stripper regularly to keep the blades sharp and free of debris. Store the tool in a dry place to prevent rust.

The wire stripper is an indispensable tool for everyone who works with electricity. Its ability to prepare wires safely and accurately makes it essential for ensuring safe and reliable electrical installations.

## Security tools:

- **Insulating gloves:** Inspect gloves before use to make sure they are not damaged.

- **Insulating mats:** Place mats under the work area and make sure they are clean and free of damage.

- **Protective goggles:** Wear goggles that provide clear vision and protection from splashes.

**Conclusions:**
The correct technique of using tools enables accurate results, minimizing the risk of injury. In the next section, we will look at the safety regulations to be followed in any work environment, ensuring a solid foundation to meet the daily challenges of the profession.

## 2.2 Safety regulations in the working environment

Safety in the work environment is of paramount importance to do-it-yourself electricians, considering the significant risks associated with working with electricity. Complying with strict safety regulations not only protects those doing the work, but also ensures that people around and property are protected. Here are some essential safety regulations that every do-it-yourself electrician should follow:

1. **Adequate training**: Before undertaking any electrical project, it is crucial to acquire proper training. Understanding the basic principles of electricity, safe working methods and emergency procedures is essential.

2. **Personal Protective Equipment (PPE)**: Always use appropriate PPE, such as insulating gloves, safety glasses, safety shoes and protective clothing. These devices minimize the risk of electric shock, burns and other injuries.

3. **Appropriate tools and implements**: Use specific tools and implements for electrical work that are insulated and meet current safety standards. Damaged or inadequate tools can significantly increase the risk of accidents.

4. **Disconnect the power supply**: Before starting any work, it is essential to disconnect the power supply. This avoids the risk of electric shock while working on the circuits.

5. **Voltage check**: Even after disconnecting the power supply, use a voltage tester to make sure the circuits are completely de-energized before starting work.

6. **Follow technical standards**: Adopt local and national regulations regarding the installation and maintenance of electrical systems. These standards are designed to ensure safe and reliable installations.

7. **Cable management**: Keep cables neat and protected to avoid tripping hazards or damage to the cable itself, which could expose electrical wires and create hazards.

8. **Labeling and documentation**: Clearly label circuits, panels, and switches to facilitate maintenance and future repairs. Maintain detailed records of modifications and installations made.

9. **First Aid**: Having a first aid kit available and knowing first aid techniques in case of electrical accidents is vital.

By following these guidelines, do-it-yourself electricians can significantly reduce the risks associated with electrical work and ensure a safe working environment for themselves and others.

## 2.3 Prevention of common electrical hazards

Prevention of electrical hazards is crucial for any technician working with electricity. Accidents can be avoided with preventive measures and a thorough knowledge of safety practices. Some of the most common threats and tips for avoiding them are listed below.

**Electric shock:** An electric shock occurs when a person accidentally touches a live conductor, allowing current to pass through the body. Shocks can range from mild to life-threatening, depending on the voltage and duration of exposure.

- **Prevention:** Make sure power is disconnected before working on any circuit. Use insulating gloves and tester to confirm the absence of voltage.

**Electric arc:** Electric arc is an explosion of energy caused by a short circuit or improper connection. This phenomenon can generate extremely high temperatures and pressure waves.

- **Prevention:** Wear fire-resistant clothing and protective goggles. Isolate and block circuits before performing maintenance and follow proper procedures for reconnection.

**Electrical fires:** Fires can be caused by overloaded cables, short circuits or malfunctioning devices.

- **Prevention:** Use cables with adequate capacity and install circuit breakers or residual current circuit breakers. Replace damaged or overheated components immediately.

**Exposure to hazardous chemicals:** Some work environments may contain corrosive or flammable chemicals that pose a risk when electrical devices are present.

- **Prevention:** Wear appropriate protective equipment and properly ventilate the work area. Ensure that electrical components are protected by appropriate enclosures.

**Falls and mechanical injuries:** Working at heights or handling heavy tools can lead to falls or injuries.

- **Prevention:** Always use certified ladders and equipment and anchor yourself with a harness if necessary. Wear protective helmets and make sure the area is free of obstacles.

**Short circuits:** Occur when two conductors at different voltages come into contact, creating a low-resistance path for current.

- **Prevention:** Check connections carefully and insulate conductors. Use disconnect switches and fuses to avoid serious damage to the circuit.

**Conclusions:** Understanding the most common hazards and how to prevent them is the first step in maintaining a safe working environment. Every technician should be able to identify potential hazards and take appropriate preventive measures.

In the next section, we will look at case studies of actual incidents and how to avoid them, providing practical examples to improve awareness and reduce errors.

## 2.4 Case studies on accidents and how to avoid them

Analyzing case studies on electrical accidents provides valuable advice on how to prevent them in the future. Through analysis of real-life situations, common mistakes and best practices can be understood. Below are some significant examples and tips for avoiding similar accidents.

**Case 1: Electric shock during maintenance** An electrician was working on repairing an electrical panel. After isolating the main power, he started to disassemble the internal components. However, he did not notice that part of the circuit was still energized, thus suffering an electric shock.

- **Causes:** The electrician had not verified the absence of voltage on all circuits and had not properly isolated the switchboard.
- **Prevention:** Use a tester to check that the entire circuit is isolated. Set up a lockout/tagout procedure to ensure that power is not accidentally restored.

**Case 2: Electrical arc during wiring** During the installation of a new system, a technician connected a live conductor to the wrong terminal. This resulted in an electrical arc, causing significant damage to the circuit and injuring the technician.

- **Causes:** The technician did not isolate the circuit during wiring and did not check the connections before mounting.
- **Prevention:** Isolate the circuit during wiring work. Check connections thoroughly before applying voltage. Wear protective clothing and use protective equipment.

**Case 3: Fire caused by overload** In one building, an old electrical system powered more devices than it could handle. The overload overheated the wiring, causing a fire that damaged part of the building.

- **Causes:** The capacity of the system was less than the required load. In addition, the protection devices were not properly installed.

- **Prevention:** Verify that installations are adequate to requirements. Use properly sized circuit breakers and fuses. Schedule periodic plant maintenance.

**Case 4: Fall Injury** A technician was working on a ladder to install an electrical conduit. During the operation, he lost his balance, falling from the ladder and sustaining serious injuries.

- **Causes:** The ladder was not properly secured and the technician was not wearing a safety harness.
- **Prevention:** Make sure ladders are stable and well positioned. Use a safety harness if working at height and make sure the area is free of obstacles.

**Conclusions:** The incidents highlight how taking preventive measures and proper training are key to reducing risks. Electricians must develop safety procedures, use appropriate tools and protective equipment, and constantly monitor conditions in their work environment.

In the next section, we will look at the design principles of home electrical systems, which are fundamental to developing safe and efficient systems in every home.

# Chapter 3: Design of Civil Installations

## 3.1 Design principles of household electrical systems

The design of home electrical systems requires attention, precision and compliance with specific technical standards. The goal is to create a system that is safe, efficient and capable of meeting the day-to-day needs of the home. Here are some key principles to follow.

**1. Electrical needs analysis:** Each home has unique electrical requirements, so the first step is to identify and assess the needs of the home and its occupants. This includes the number of appliances to be connected, the presence of high-powered appliances (such as electric furnaces and air conditioners), and the distribution of outlets in each room. How is this done? To analyze the electrical needs of a home with a view to creating an electrical system, it is essential to evaluate several factors:

1. **Determining the total load**: Calculate the expected electrical load by adding up the consumption of all the appliances and devices that will be used.

2. **Space** arrangement: Consider the layout of rooms and the functionality required in each area to appropriately place outlets and switches.

3. **Safety regulations**: Ensure that the project complies with all local regulations related to electrical safety and building codes.

4. **Future expansions**: Provide for any future expansions or changes in the use of living space to avoid complex renovations to the facility.

5. **Energy efficiency**: Integrate energy-saving and sustainability solutions, such as the use of energy-efficient devices and the installation of energy management systems.

**2. Planning the electrical panelboard:** The electrical panelboard is the heart of the system. It should be sized to support present and future loads and facilitate expansion or modification. Organizing circuits logically, such as separating outlets in different rooms from lights, helps reduce risks and improve load management.

**3. Circuit sizing:** A properly sized circuit minimizes overload risks and ensures safety. Cables should be selected according to the maximum expected current and distance from the source. It is also critical to correctly calculate the number of outlets for each circuit to avoid overloading and waste.

**4. Protection and grounding:** Every installation must be equipped with protections such as earth leakage circuit breakers (circuit breakers) and automatic circuit breakers. Grounding is essential to avoid electrocution hazards. Every outlet and every metal component connected to the installation should be properly connected to the grounding system.

**5. Strategic placement of outlets:** The location of outlets should be designed to be convenient and meet the needs of users. Each room should have an adequate number of outlets, placed near areas where the most use is expected, such as corners for electronics, work or rest areas.

**6. Lighting**: Lighting system design should consider both functionality and energy efficiency. The use of dimmer switches, motion detectors, and energy-efficient lamps can improve the experience and reduce costs. Distributing lights evenly ensures good lighting in all areas of the home.

**7. Home automation and control:** Including automation systems makes the system smarter and more flexible. Controlling lights, appliances, and security systems via smartphones or voice assistants improves convenience and saves energy.

In the next section, we will look at current regulations and compliance needed to ensure that the facility is safe and meets legal standards.

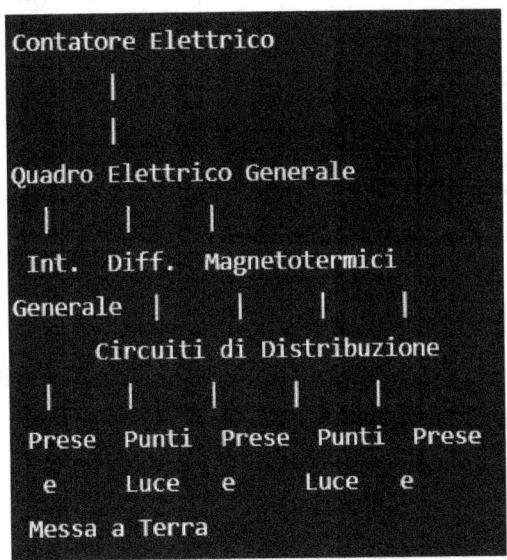

## 3.2 Plant planning and layout

Planning and layout of electrical installations are crucial to ensure that a system is functional, safe, and meets technical standards. A well-designed layout allows electricians to perform the installation more efficiently and simplifies future maintenance.

1. **Environment and needs analysis:** Before developing the layout, it is essential to conduct an analysis of the environment in which the system will be installed. Identify the specific needs of the building (residential or commercial), the layout of rooms, and the location of energy-intensive devices (such as furnaces, air conditioners, and servers).

2. **Determination of circuits:**

    - **Load allocation:** Distribute loads among different circuits to avoid overloading and ensure a stable supply. For example, lights and outlets should be on separate circuits, while high-wattage appliances (washing machines, ovens) require dedicated circuits.

    - **Load calculation:** To calculate the load of a household electrical panel, sum the maximum power of all the appliances that will be used at the same time. Use the formula $P = V \times I$, where P is the power in watts, V is the voltage in volts, and I is the current in amperes. This helps you determine the appropriate size of the circuit breakers and the panelboard itself.

3. **Electrical panel layout:** The electrical panel should be easily accessible, ideally near the building entrance or in a service room. It should include a main switch and disconnect switches for each circuit.

    - **Neat arrangement:** Make sure circuits are clearly identified with labels.

- **Differential protection:** Each circuit group should have dedicated differential protection, and the switchboard should be equipped with a general protection device.

3. **Cable distribution:** When distributing cables in a household electrical system, it is essential to consider both safety and functionality. Cables should be placed inside protective tubes (conduits) or wall ducts, following routes that avoid damp and excessive heat zones. Use PVC-insulated copper cables, suitable for household use. For sockets, generally use 2.5 mm² cables, for lighting 1.5 mm² cables. Be sure to comply with current electrical regulations (e.g., IEC) to ensure the safety and efficiency of the installation.

5. **Placement of sockets and switches:**

    - **Standard height:** Place sockets and switches at a standard height that allows comfortable use.
    - **Distribution:** Make sure each room has an adequate number of outlets, especially near work areas such as the kitchen or electronic corners.
    - **Lighting:** Provide a sufficient number of switches for lighting, with controls near entrances and in the busiest areas.

6. **Drafting the project:** Once all details have been established, draft a project that includes:

    - **Diagrams and diagrams:** Draw detailed layouts of circuits and cable routes.

- **Materials list:** Provide a complete list of components and materials needed.

**Conclusions:** Planning and layout are essential to designing an installation that is safe, functional, and efficient. A detailed plan allows electricians to follow a clear path during installation and maintenance.

In the next section, we will look at how to select materials and components, ensuring that they meet safety standards and plant requirements.

## 3.3 Choice of materials and components

The choice of materials and components is crucial to ensuring that a home electrical system is safe, efficient, and able to meet consumer needs. By choosing carefully, you can not only comply with regulations, but also reduce the risk of failures and malfunctions.

**1. Cables and conductors:**

- **Cross-section selection:** The cross-section of the cables should be chosen according to the expected current for each circuit. In general, circuits with higher current draw, such as those for high-power appliances, require cables with larger cross-sections.

- **Conductor Material:** Copper conductors are the most widely used because of their conductivity and strength. Aluminum, which is cheaper, is less conductive and requires special connectors.

- **Insulation:** The insulation material must be able to withstand the temperature and environment in which it is used. In domestic cables, PVC insulation is common, while rubber materials are preferred in places with high temperatures.

## 2. Ducting and piping:

- **Material:** Plastic or metal cable trays should be chosen according to the environment in which they are installed. Plastic is ideal for indoors, while metal withstands shocks and outdoor elements better.

- **Size: They** should be roomy enough to comfortably hold cables while retaining space for future expansion.

- **Fixing system:** Wireways should be firmly fixed and sealed to prevent dust and moisture accumulation.

## 3. Protective devices:

- **Residual current circuit breakers (RCDs):** These devices are essential to prevent electrocution and short circuits. Sensitivity should be appropriate for the specific application, usually 30 but for household use.

- **Circuit breakers:** Protect circuits from overloads and short circuits. The rated current should be selected according to the circuit capacity.

- Fuses: Fuses are safety devices in an electrical system that protect against overloads and short circuits by interrupting the flow of current through a melting filament. They provide an extra level of protection and should be installed in critical circuits, such as high-power circuits.

## 4. Sockets and switches:

**Sockets**: Use sockets with circuit breakers to protect against overloads and short circuits. Choose sockets appropriate to CEI standards, suitable for the type of electrical load and the required degree of protection (e.g., IP44 for wet environments).

**Cables**: Copper cables are the most common because of their conductivity and durability. For general loads, use PVC cables with 2.5 mm² cross-section for outlets and 1.5 mm² for lighting. Make sure the insulation is suitable for the installation environment, such as reinforced conduit for outdoor passages or corrosive environments.

Circuit breakers are crucial components of an electrical system, used to control, protect and isolate electrical circuits. Common materials include heat-resistant plastic and metal components to ensure durability and safety.

**Types of switches:**

1. **Circuit breakers**: They protect against overloads and short circuits, automatically shutting down the circuit in case of faults.

2. **Residual** current **circuit breakers**: They detect current leakage to ground, immediately shutting off the power supply to prevent electric shock.

**Selection Criteria:** Choose circuit breakers based on circuit current rating, type of load, and sensitivity required for leakage current protection. Ensure that the circuit breakers meet local and international regulations for electrical system safety.

## 5. Lighting fixtures:

- **Efficiency:** Fixtures should allow the use of energy-efficient bulbs such as LEDs or CFLs.
- **Protection:** Like sockets, they must have IP protection appropriate for the installation environment.
- **Aesthetics and functionality:** In addition to efficiency, they must be strategically placed for optimal lighting.

In the next section, we will provide practical examples of home circuit design to illustrate how to implement the principles discussed and choose the most suitable components for each need.

## 3.4 Domestic circuit design examples

The design of home circuits requires attention to detail, a thorough knowledge of current regulations, and the use of appropriate materials and tools to ensure safety and efficiency. The following are practical examples of circuits typically installed in a home environment, complete with detailed instructions and specifications on the materials to be used.

### Example 1: Installation of a circuit for indoor lighting

**Materials needed:**

- Insulated copper electrical cable (1.5 mm² for lighting)
- Wall switches
- Ceiling lights or other lighting fixtures
- Junction boxes
- Terminal blocks

**Tools needed:**

- Screwdriver
- Cutter
- Wire stripper
- Voltage tester

**Steps:**

1. **Route design**: Define the route of cables from the electrical panel to the various light sources, avoiding areas of high humidity and heat sources.

2. **Cable installation**: Place cables inside protective conduits attached to the wall. Make sure each section of cable is well anchored and protected.

3. **Connecting the switches**: Install the switches in the ready-made flush-mounted boxes. Connect the wires to the terminals of the switches using the terminals.

4. **Mounting the ceiling lights**: Attach the ceiling lights to the ceiling or wall and connect the electrical wires to the corresponding terminals on the light fixture.

5. **Final test**: Verify the integrity of the circuit using a voltage tester before reactivating the general power supply.

## Example 2: Installation of electrical outlets

**Materials needed:**

- Insulated copper cable (2.5 mm² for sockets)
- Recessed electrical outlets

- Recessed boxes for sockets

**Tools needed:**

- Screwdriver
- Drill with masonry bit
- Spirit level
- Voltage tester

**Steps:**

1. **Defining the location of outlets:** Choose the location of outlets according to your usage needs, considering furniture layout and ease of access.

2. **Installation of junction boxes:** Use the drill to create recesses in the wall where you will install the recessed boxes.

3. **Laying** cables: Lay the cables from the electrical panel to the flush-mounted boxes using protective conduits. Secure cables with appropriate clamps.

4. **Connecting the sockets:** Connect the wires to the socket terminals, observing the indicated polarity. Fix the sockets in the flush-mounted boxes.

5. **Verification and set-up:** Check the connections for correctness with a voltage tester. Once verified, restore power and test the outlets.

**Example 3: Repair of a furnace with burn marks due to electricity**

Repairing an electric furnace that shows signs of burning requires attention to detail, precision and proper safety measures. Here's how to proceed step by step:

**Materials needed:**

- High temperature electrical cables
- Heat-resistant electrical connectors
- Insulating tape for high temperature

**Tools needed:**

- Screwdriver
- Pliers
- Multimeter
- Flashlight

**Steps for repair:**

**Step 1: Disconnecting the power supply**

- First, make sure the oven is disconnected from the power supply to avoid the risk of electric shock.

**Step 2: Visual inspection**

- Open the back or bottom panel of the oven using a screwdriver. Use a flashlight to inspect the interior for obvious burn marks on wiring and electrical components.

**Step 3: Diagnosis**

- Use a multimeter to test the continuity of electrical wires. Check for breaks in the circuit that could indicate burned or faulty wires.

- Check connectors and terminals for signs of corrosion or heat damage.

## Step 4: Replacement of damaged components

- Carefully remove damaged components, such as burned-out cables or corroded connectors. Replace them with new, high-temperature-specific cables and connectors, making sure to observe the original polarities and connections.

- Securely fasten the new cables with heat-resistant electrical connectors and make sure there are no exposed parts.

## Step 5: Isolation and verification

- Use insulating tape suitable for high temperatures to insulate all repaired connections.

- Re-check all connections to make sure they are secure and well insulated.

## Step 6: Final test

- Reconnect the oven to the power supply and turn it on for a brief test. Check that all oven functions are operational and that there are no signs of smoke or burning smell.

## Step 7: Re-assembly

- Once you have confirmed that the oven is working properly, put the back or bottom panel back in place and make sure all screws are tight.

**Conclusion and precautions:**

- Keep the oven and surrounding area clean and free of flammable materials. This not only extends the life of the oven but also reduces the risk of future electrical problems.

These examples illustrate basic steps for designing and installing circuits for lighting and outlets in a home environment, ensuring a smooth transition to the next chapter on safe system installation techniques.

# Chapter 4: Installation of Installations

## 4.1 Steps for safe installation of equipment

Safe installation of electrical systems in a home requires not only specific technical skills, but also strict adherence to safety regulations to protect both installers and end users from electricity. Below, the basic steps and tools needed to properly install an electrical system are detailed, ensuring a smooth transition to advanced wiring and connection techniques.

**Tools needed:**

- Insulated screwdriver
- Insulated pliers
- Cutter
- Wire stripper
- Voltage tester
- Spirit level
- Electric drill
- Metal hacksaw or miter saw

**Steps for installing an electrical system:**

**Step 1: Design and planning**

- Begin with a detailed design of the electrical system. Define the locations where outlets, switches, and equipment will be placed. Use CAD software or detailed schematics to accurately plan the routing of cables and the location of system elements.

## Step 2: Installation of junction boxes

- Mount the junction boxes (connect and protect electrical wires) at the designated locations. Make sure they are securely attached to the wall or ceiling and leveled carefully. These boxes will serve as connection points for switches, outlets and other devices.

## Step 3: Laying the cables

- Lay cables following the routes provided in the design phase. Use protective conduits or corrugated tubing to protect cables and ensure they are laid away from sources of heat and moisture. Secure cables with cable clips to prevent them from moving or bending.

## Step 4: Connecting sockets and switches

- Once all the wires are laid, proceed to connect the outlets and switches. Use the voltage tester to make sure there is no current while working. Connect the wires into the appropriate terminals following the standard wiring diagrams.

## Step 5: Installation of equipment and accessories

- Install necessary equipment such as lamps, heaters or other electrical devices. Connect each device to the previously laid cables, making sure to follow the manufacturer's instructions to avoid connection errors.

## Step 6: Verification and testing of the system

- After completing all connections, check the system using the voltage tester to make sure everything is connected correctly and working as intended. Check that there are no power leaks and that all devices are operating properly.

**Step 7: Documentation and labeling**

- Complete the installation by labeling all circuits in the electrical panel and making detailed documentation of the installation. This will be essential for future maintenance and for any upgrades or repairs.

**Practical example:** Suppose you need to install electrical wiring in a new kitchen. After planning the location of outlets and switches, assemble the junction boxes and lay the cables in the predefined routes. He then installs the necessary outlets for major appliances such as the refrigerator and oven, as well as switches for lighting. Finally, he conducts a series of tests to check the effectiveness and safety of the installation.

These steps not only ensure safe and compliant installation, but also facilitate the transition to the next point, which deals with wiring and connection techniques, ensuring a complete understanding of how to effectively manage a home electrical project from start to finish.

## 4.2 Wiring and connection techniques

The installation of effective and safe wiring, is essential to the functionality of any home electrical system. Wiring and connection techniques must ensure reliability, accessibility, and safety. This chapter provides detailed guidance on how to perform these tasks correctly, setting the stage for integration with home automation systems and smart devices.

## Tools needed:

- Cutter
- Wire stripper
- Insulated screwdrivers
- Crimping tool
- Voltage tester
- Insulating tape
- Labeler

## Steps for wiring and connection techniques:

### Step 1: Cable route planning

- Before starting, it is crucial to plan the cable route within the home. Consider access points, distance from heat sources, and ease of maintenance. Use floor plans of the house to define the most efficient route.

### Step 2: Cable installation

- Use ducts or conduits to protect cables and keep them neat. Secure conduits to walls or ceilings with screws and dowels, taking care not to damage cables with excessive pressure.

### Step 3: Cable preparation

- Cut the wires to the desired length with a wire cutter. Use a wire stripper to remove the insulation without damaging the internal wires. For connections, strip about 1 cm of insulation from the terminal of each wire.

## Step 4: Connecting the cables

- Connect wires to junction boxes, switches, outlets, and other devices using appropriate terminals. Make sure each connection is secure and well insulated with electrical tape or heat shrink tubing.

## Step 5: Test the wiring

- Once the connections are completed, use a voltage tester to check for shorts, overloads, or other problems. Check each circuit to make sure it is working properly.

## Step 6: Labeling and documentation

- Label each circuit in the electrical panel and note all changes in the house records. This will simplify future maintenance and possible troubleshooting.

## Practical example: Installation of electrical outlets

- **Materials**: Copper cables (2.5 mm$^2$), electrical sockets, flush-mounted boxes, screws.
- **Tools**: Wire cutter, wire stripper, screwdriver, voltage tester, electrical tape, spirit level.
- **Steps:**
    1. **Plan the location of the sockets**: Use a level to ensure that the flush boxes are installed level.
    2. **Install the** cables: Lay the cables from the main circuits to the locations of the new outlets following the conduits.

3. **Prepare the** cables: Strip the cables and prepare the terminals for connection.

4. **Connect the sockets**: Fix the wires into the corresponding terminals of the sockets, following the wiring diagram.

5. **Test the connections**: Check that all outlets are working properly without signs of overheating or sparks.

These techniques not only ensure safe and functional installation of the electrical system, but also facilitate integration with advanced systems such as home automation and smart devices, topics discussed in Section 4.3 below.

## 4.3 Home automation and intelligent systems

The integration of home automation and smart systems represents a quantum leap in the efficiency, comfort, and safety of modern homes. These systems allow users to manage lighting, air conditioning, security, and other home devices via voice commands or apps on smart devices. In the following, how to implement a home automation system is described, preparing the transition to verification and testing of installed systems (Section 4.4).

**Tools needed:**

- Smartphone or tablet to configure devices
- Screwdrivers
- Voltage tester
- Multimeter
- Drill and assembly tools

**Steps for installing a lighting automation system:**

1. **Device selection**: Choose devices that are compatible with your home network and can be integrated with other smart systems already in place or planned for the future. Opt for smart bulbs and switches that support control via apps and voice assistants such as Google Assistant, Amazon Alexa or Apple HomeKit.

2. **Physical installation**:

    - **Replacing traditional bulbs**: Replace existing bulbs with smart bulbs. Make sure the power is turned off during this step to avoid electrical hazards.

    - **Smart switch installation**: Mount the smart switches in the pre-existing light points. Use the drill to secure the switches and the screwdriver to connect the wires to the switches, following the manufacturer's instructions.

3. **Network configuration**:

    - **Connecting devices to your home Wi-Fi network**: Follow the manufacturer's instructions to connect each device to your home network via the dedicated app.

    - **Software updates**: Be sure to install the latest available software updates for your smart devices, thereby improving functionality and security.

4. **Programming and customization**:

- **Setting Scenarios**: Configure scenarios and routines to automate daily tasks, such as turning off all lights when you leave the house or turning on some lamps when you return.
- **Integration with other devices**: Connect the lighting system with other smart devices, such as thermostats or security systems, for integrated and automated home management.

5. **System testing**:
    - **Check operation**: Test each component to make sure it responds correctly to app and voice commands.
    - **Interoperability check**: Make sure all devices are communicating properly with each other and with the control unit.

## Practical example: Living room lighting automation

- **Materials**: smart bulbs, smart switches.
- **Tools**: Screwdriver, voltage tester, smartphone.
- **Steps**:
    1. Turn off the power and replace existing bulbs and switches.
    2. Connect the new bulbs and switches to the Wi-Fi network via the dedicated app.
    3. Configure custom routines to automate the turning on and off of lights.

4. Test the system to confirm correct installation and configuration.

These steps not only improve the comfort and efficiency of the home, but also prepare the system for detailed verification and testing, ensuring that everything works as intended and that the system is safe and reliable, as discussed in the next section 4.4 of the book.

## 4.4 Verification and testing of installed systems

The process of checking and testing installed systems is essential to ensure that all components of the electrical system are functioning properly and safely. This step allows any installation errors or defective components to be identified before the system is permanently put into service.

**Tools needed:**

- Multimeter
- Insulation tester
- Continuity tester
- Current clamp
- Thermal imaging camera (optional for thermal checks)

**Steps for testing an electrical system:**

**Step 1: Preparation**

- Make sure that the whole system is turned off before starting testing. This includes turning off the main power supply to avoid risks during testing.

## Step 2: Verification of circuit continuity

- Use a continuity tester to make sure all circuits are complete and have no interruptions. Check that connections in electrical panels and terminations to devices are firm and correct.

## Step 3: Measurement of insulation resistance

- With an insulation tester, measure the insulation resistance of cables and equipment. Values that are too low may indicate insulation problems that could lead to short circuits or faults.

## Step 4: Checking differential protection

- Test residual current circuit breakers to ensure that they are working properly. This is vital to prevent accidents due to current leakage to ground.

## Step 5: Verification of the functionality of protective devices

- Check that all protective devices, such as circuit breakers, are properly calibrated and react as intended under load.

## Step 6: Thermographic control (optional)

- If available, use a thermal imaging camera to identify abnormal hot spots that may indicate overloads, bad connections, or other anomalies that are not visible to the naked eye.

## Step 7: Operational test

- Turns the power back on and operates each component of the system to verify proper functionality. This includes turning on lights, testing electrical outlets and other devices connected to the system.

**Practical example: Testing a new lighting circuit**

- **Materials**: Multimeter, continuity tester, current clamp.
- **Steps**:
    1. **Continuity**: Check the continuity of the wires from the electrical panel to the switches and from the lights.
    2. **Insulation**: Measures the insulation resistance of lighting cables.
    3. **Load**: Turn on the lights and observe the behavior of the circuit under load, checking the operation of the switches and the absence of voltage variations.
    4. **Final verification**: Use the current clamp to measure the current flow and confirm that it matches the technical specifications.

This testing process not only ensures that the facility is safe and functional when it is activated, but also provides a solid foundation for maintaining efficiency and resolving common problems that may arise during continued use, a topic covered in section 4.5 of the book below.

## 4.5 Common troubleshooting during installation

During the installation of an electrical system, several common problems can arise that require immediate attention to ensure the safety and functionality of the installation. Identifying and resolving these problems is essential to prevent future failures and to maintain the efficiency of the installation. This chapter explores strategies for dealing with these problems, integrating seamlessly with the continuation of the book that deals with ongoing operational troubleshooting.

## Tools needed:

- Voltage tester
- Multimeter
- Current clamp
- Insulated screwdriver
- Thermal imaging camera (optional for overload detection)

## Steps to solve common problems during installation:

### Step 1: Initial Diagnosis

- Before proceeding, use a voltage tester to make sure that the power is off in the work area. This step is critical to avoid risks during diagnosis.

### Step 2: Verification of connections

- Check all electrical connections to make sure they are firm and properly insulated. A loose connection can cause intermittent interruptions and potential hazards.

### Step 3: Current measurement

- Use a current clamp to measure the current flowing through circuits. This will help you identify overloads or abnormalities in current flow.

### Step 4: Checking switches and fuses

- Examine circuit breakers and fuses for damage or burn out. Replace them if necessary to ensure protection against overloads and short circuits.

### Step 5: Using the multimeter

- Set the multimeter to measure resistance or continuity to check for open or short circuits in the wires.

### Step 6: Thermographic inspection

- If available, use a thermal imaging camera to identify hot spots. These may indicate areas of potential overload or insufficient insulation.

### Practical example: Solving a light intermittency problem

- **Tools needed**: Voltage tester, insulated screwdriver, multimeter.

- **Steps**:

    1. **Circuit isolation**: Make sure the affected circuit is isolated and safe to work on.

    2. **Checking the switch**: Check the light switch for signs of wear or damage. Use the voltage tester to test the integrity of the switch.

    3. **Checking connections**: Open the switch box and use the screwdriver to tighten all connections.

    4. **Multimeter measurement**: Set the multimeter to continuity to test the connection between the switch and the light.

5. **Power restoration and test**: Turn the power back on and check that the light is working properly without flickering.

These steps not only help solve common problems during installation, but also establish a solid foundation for effective ongoing maintenance and troubleshooting, as will be explored in section 5.1 of the book below.

# Chapter 5: Maintenance and Control

## 5.1 Troubleshooting common problems during installation

During the installation of electrical systems, various problems may be encountered that require quick diagnosis and precise solutions. An understanding of common problems allows them to be solved effectively, ensuring a functional and safe installation.

**1. Faulty connections:**

- **Causes:** Faulty connections are often caused by insufficient tightening, oxidation, wear, or improper connection techniques.

- **Solution:** Check each connection using a continuity tester. Ensure that the wires are firmly seated in the terminals or screws. Clean oxidized connections with sandpaper or replace them. Always choose quality terminal blocks and cables with proper sheathing.

**2. Short circuits and overloads:**

- **Causes:** A short circuit can be caused by exposed wires, loose connections, or overloaded cables. Overloads occur when too many devices are connected to a single circuit, exceeding the capacity of the circuit breaker.

- **Solution:** Isolate each circuit and check each segment with a multimeter. Identify points where resistance is zero and solve by removing loose connections or replacing wires. Distribute loads evenly and add new circuits if necessary.

## 3. Inadequate wiring:

- **Causes:** Inadequate wiring may result from cables that are too thin for the intended load or unprotected cable runs, causing overheating or mechanical damage.

- **Solution:** Check the cross-section of the cables used against the power drawn by each device and the distance covered. Ensure that cables are well protected by conduit or sheathing and replace damaged ones.

## 4. Faulty circuit breakers:

- **Causes:** Circuit breakers may be defective due to poor manufacture, repeated overloading or incorrect installation.

- **Solution:** Test circuit breakers by removing them from the circuit and seeing if they trip properly. If a circuit breaker continues to trip for no apparent reason, it may be defective and should be replaced.

## 5. Problems with lighting:

- **Causes:** Lighting systems may have problems with flickering, intermittence or failure to light due to faulty bulbs, unstable voltage, loose connections or worn switches.

- **Solution:** Check the voltage in the circuit and test each bulb. Replace the defective bulbs and make sure the wiring is correct. If the problem persists, you may need to replace the switch or use a more stable power supply.

6. **Noise in devices:**

- **Causes:** Unusual noises in electrical devices can be caused by overloaded transformer, loose wiring or vibration in electrical panels.

- **Solution:** Check the load on transformers, replacing or redistributing circuits if necessary. Tighten connections in electrical panels and secure vibrating components with bolts or brackets.

**Conclusions:** Timely troubleshooting of common problems during installation is essential to ensure that the system functions properly and is safe. Methodical testing and proper circuit planning can help prevent these problems.

In the next section, we will look at techniques for inspecting systems to keep them in optimal condition and detect hidden problems before they become critical.

## 5.2 Plant inspection techniques

Regular inspection of electrical systems is crucial to ensuring the safety, efficiency, and proper operation of a home. Implementing systematic inspection techniques helps prevent failures, reduce fire hazards, and ensure compliance with applicable regulations.

**Tools needed:**

- Voltage tester
- Multimeter
- Thermal imaging camera
- Current clamp
- Insulation tester

## Steps for inspecting an electrical system:

### Step 1: Planning the inspection

- Schedule the inspection during a period when the facility may be temporarily disabled, ensuring that all procedures are safe for those performing the inspection.

### Step 2: Visual inspection

- Begin with a visual inspection of all accessible components of the system, such as electrical panels, outlets, switches, and exposed wiring. Look for signs of wear and tear, physical damage, overheating, and corrosion.

### Step 3: Voltage and continuity test

- Use a voltage tester to check for the absence of voltage before proceeding with other measurements. Then use a multimeter to test the circuits for continuity. This step is essential to identify any interruptions or shorts.

### Step 4: Measurement of insulation resistance

- Employs an insulation tester to measure the insulation resistance of cables and other electrical components. This helps identify whether the insulation is still effective or has been compromised.

### Step 5: Checking the connections

- Check all electrical connections to make sure they are tight and there are no signs of oxidation or relaxation that could affect the safety and efficiency of the system.

## Step 6: Load measurement

- Use a current clamp to measure the current flowing through major circuits to ensure they do not exceed design specifications, indicating potential overloads.

## Step 7: Thermal sensing

- Employs a thermal imaging camera to identify hot spots that may indicate hidden problems such as poor connections, overloads, or appliance failures.

## Practical example: Inspection of electrical panel

- **Tools needed**: Voltage tester, multimeter, thermal imaging camera, current clamp, insulation tester.

- **Steps**:

    1. **Turn off the electrical panel**: Make sure there is no voltage using the voltage tester.

    2. **Visual inspection of** the panel: Examine the panel for signs of damage or wear.

    3. **Circuit continuity test**: Use the multimeter to ensure that all circuits in the cabinet are intact.

    4. **Insulation resistance measurement**: Verifies the integrity of the insulation of cables connected to the switchboard.

    5. **Thermal sensing**: Use thermal imaging to check for overheating in components.

6. **Final verification**: Reconnect the switchboard, turn on the voltage, and monitor the system response.

This systematic approach not only ensures that plant functionality and safety is maintained but also facilitates the transition to more sophisticated monitoring and diagnostic techniques.

## 5.3 Technical documentation management

Effective management of technical documentation is crucial to the long-term maintenance and safety of a home's electrical systems. Every detail, from design to maintenance, must be carefully documented to ensure that the necessary information is available for future inspections, repairs, and upgrades. This practice not only helps preserve compliance with current regulations but also makes the work of technicians and engineers easier over time.

Technical documentation should include:

- **Detailed wiring** diagrams: Updated diagrams representing the current configuration of the installation, including all changes made since the original installation.

- **Material specifications**: A complete list of materials used, including types of cables, protective devices, and control equipment, with details on their manufacturers and technical specifications.

- **Installation records**: Detailed notes on how and when system components were installed or modified, who performed the work, and the procedures adopted.

- **Inspection and test results**: Reports on function and safety tests, including insulation resistance testing, circuit continuity testing, and differential testing.
- **Device Manuals**: Technical manuals and operating guides for each piece of installed equipment, providing valuable guidance for maintenance and troubleshooting.

**Practical example:**

Suppose that the lighting system of a residence was upgraded by integrating automation elements. Documentation of this intervention should include:

- Updates to wiring diagrams to reflect new connections and configurations.
- An inventory of newly installed components, including smart switch models and types of LED bulbs used.
- A detailed history of the installation, with notes on any challenges encountered and solutions adopted.
- Results of post-installation functionality tests to ensure that everything works as expected.

This information becomes critical for comparing past and present maintenance practices and for planning future interventions, issues that will be further explored in the next section, devoted to case studies of effective maintenance.

## 5.4 Case studies of effective maintenance

Effective maintenance of household electrical systems is critical to ensure safety, energy efficiency and long life of the systems. Through three practical case studies, this chapter illustrates how to address and solve common problems, using appropriate tools and established methodologies. These examples will help to better understand maintenance practices, setting the stage for Section 6.1 below, which discusses methodologies for fault identification.

## Example 1: Replacement of a Defective Switch

**Tools needed:**

- Voltage tester
- Screwdriver
- Clamp

**Materials:**

- New electrical switch

**Steps:**

1. **Turn off the power:** Make sure the power is off to avoid electrical hazards.
2. **Check for voltage:** Use the voltage tester to confirm that the switch is current-free.
3. **Removing the faulty switch:** Remove the switch cover and use the screwdriver to disconnect the wires.
4. **Installation of the new switch:** Connect the wires to the new switch following the manufacturer's directions.
5. **Test the new switch:** Restore power and check the operation of the switch.

## Example 2: Repairing an Electrical Outlet

**Tools needed:**

- Multimeter
- Screwdriver

**Materials:**

- New electrical outlet

**Steps:**

1. **Safety:** Turns off power from the electrical panel.
2. **Checking the socket:** Use the multimeter to check for voltage.
3. **Socket replacement:** Disconnect the wires from the damaged socket and connect the new socket.
4. **Verify:** Reset the power supply and test the socket to ensure that it is working properly.

**Example 3: Correcting an Overload Problem**

**Tools needed:**

- Current clamp
- Screwdriver

**Materials:**

- None specific, control of existing plant

**Steps:**

1. **Overloaded circuit identification:** Use the current clamp to measure current at various points in the system.
2. **Load redistribution:** Redistribute some devices over several circuits to reduce the load on a single circuit.

3. **Circuit verification:** After redistribution, measure the current again to make sure the levels are within safe limits.

These case studies not only show how to handle common situations but also how systematic steps and proper use of tools can prevent future problems, ensuring safe and reliable plant operation. The lessons learned here form a solid foundation for the next section, which discusses advanced fault identification methodologies that are essential for diagnosing and solving more complex problems.

# Chapter 6: Fault Diagnosis

## 6.1 Methodologies for fault identification.

Fault identification in electrical systems requires a methodical approach to ensure accurate diagnosis and effective repair. By following these methodologies, complex problems can be identified and solved while minimizing downtime and safety risks.

1. **Understanding of symptoms:**

   - **Observation:** Listen to the users' account of the symptoms they experience (e.g., outages, flickering of lights, tripping of switches) and directly observe any abnormal behavior of the system.

   - **Contextualization:** Understand where in the system problems occur and under what conditions (e.g., when turning on certain appliances or during peak loads).

2. **Preliminary visual examination:**

   - **Electrical panel:** Check that the electrical panel is well organized, with no obvious signs of overheating, exposed cables or loose connections.

   - **Connections and cables:** Check connections in devices and condition of cables, looking for signs of burns, wear, or damaged insulation.

   - **Components:** Inspect components such as switches, sockets, and protective devices for visible damage.

3. **Electrical measurements:**

- **Continuity:** Use a tester to measure circuit continuity and identify breaks in wiring.

- **Voltage and current:** Measure voltage and current at different points in the circuit to detect surges, drops or overloads.

- **Insulation:** Use an insulation meter to check the integrity of cables and make sure there is no leakage between conductors and ground.

### 4. Root cause identification:

- **Faulty circuit:** If the fault is associated with a particular circuit, isolate the segments one at a time to determine where the problem lies. This may include disconnecting loads to rule out problems with faulty appliances.

- **Components:** Individually test components such as switches, outlets, and protective devices to determine if any of them are defective.

- **Overloads:** Evaluate the load on each circuit and compare it with its capacity. An overload could cause tripping of circuit breakers or differential breakers.

### 5. Troubleshooting:

- **Repair:** Once the root cause has been identified, make necessary repairs such as tightening connections, replacing defective components, or redoing damaged wiring.

- **Upgrades:** If the failure is due to overload or obsolete design, consider upgrading circuits to meet current requirements.
- **Test:** After repair, test the circuit again to make sure that the fault has been solved and that there are no residual problems.

**Conclusions:** Fault identification methodologies are based on a systematic examination of symptoms and components. Following these steps ensures accurate diagnosis and allows problems to be addressed effectively, preventing future failures.

In the next section, we will explore the use of the multimeter and oscilloscope, key tools for making detailed diagnoses and accurate measurements in electrical systems.

## 6.2 Fault tracing and mapping techniques.

Accurate fault tracing and mapping is essential to effectively diagnose and resolve problems in a home electrical system. This process not only helps to locate the fault, but also to understand the building's electrical network, facilitating future interventions. In this segment, we will illustrate how to apply these techniques through three practical examples, providing a smooth transition, dealing with the repair of frequent faults.

### Example 1: Identification of a Fault in a Lighting Circuit.

**Tools needed:**

- Voltage tester
- Multimeter
- Flashlight

**Materials:**

- Replacement electrical cables (if necessary)
- Electrical connectors

**Steps:**

1. **Circuit isolation:** Disconnect the circuit from the main switch to ensure safety.
2. **Voltage check:** Use a voltage tester to confirm that there is no current in the circuit.
3. **Visual inspection:** Inspect the circuit using a flashlight to look for visible signs of damage such as burned or deteriorated wires.
4. **Continuity measurement:** Using the multimeter, check the continuity of cables to identify any breaks.
5. **Repair or replace:** Replace damaged wires and reconnect everything following the circuit diagrams.

**Example 2: Locating a Fault in an Electrical Outlet**

**Tools needed:**

- Multimeter
- Screwdriver

**Materials:**

- New electrical outlet (if necessary)

**Steps:**

1. **Power disconnection:** Make sure the power is off in the electrical panel.
2. **Socket test:** Use a multimeter to test the socket and see if there is continuity and correct voltage.
3. **Socket inspection:** Open the socket and check for loose or oxidized wires.
4. **Repair or replace:** Replace the socket if damaged or rearrange the wires and tighten them properly.

**Example 3: Tracking a Fault in a Heating System**

**Tools needed:**

- Thermal imaging camera
- Multimeter

**Materials:**

- Spare parts for the heater

**Steps:**

1. **Initial check:** Check the thermostat display for errors or warnings.
2. **Using the thermal imaging camera:** Identifies abnormal hot spots in the circuits or heater.
3. **Checking electrical components:** Use multimeter to measure resistances and voltages in heater components.
4. **Repair:** Replace defective electrical components and verify that the system is working properly after service.

These examples illustrate how advanced tracking and mapping techniques can facilitate diagnosis and troubleshooting in a variety of domestic settings. Understanding and applying these methods not only solves current problems but also establishes a solid foundation for preventive maintenance and effective management of future failures.

## 6.3 Repair of frequent faults

Repairing frequent faults in electrical systems requires specific skills to deal with common situations, such as short circuits, overloads, and faulty connections. Accurate diagnosis and precise repair help prevent recurring problems and ensure the safe operation of the system.

### 1. Short circuits:

- **Causes:** Short circuits are caused by exposed conductors or damaged wires coming into contact with each other, generating a high current that can overheat components or trip breakers.

- **Repair:**

    - Isolate the circuit involved and identify the location of the short circuit by a continuity test or by measuring resistance.
    - Repair or replace damaged cables, making sure they are properly insulated.
    - Check all connections to make sure they are firm and well protected.

2. **Overloads:**

    - **Causes:** Overloads occur when the total load connected to a circuit exceeds its capacity, tripping circuit breakers or differential switches.

    - **Repair:**

        - Identify the devices connected to the circuit and measure the total current drawn.

        - Distribute loads among multiple circuits, adding new circuits if necessary.

        - Replace circuit breakers with higher-capacity models if wiring permits.

3. **Faulty connections:**

    - **Causes:** Faulty connections are often due to insufficient tightening, oxidation, damaged insulation, or inadequate connections.

    - **Repair:**

        - Check each connection, tightening terminals and cleaning oxidized surfaces.

        - Replace worn terminals or damaged wiring.

        - Ensure that each connection is well insulated and free of bare conductors.

4. **Voltage drops:**

    - **Causes:** Voltage drops can be caused by cables of inadequate cross-section, excessively long runs, or loose connections.

    - **Repair:**

- Check cable cross-sections against route length and load. Replace cables with larger cross sections if necessary.

- Check the connections to make sure they are tight.

- Reduce path lengths if possible or distribute loads on separate circuits.

## 5. Defective protective devices:

- **Causes:** Protective devices, such as differential and circuit breakers, can become defective due to wear and tear or repeated tripping.

- **Repair:**

    - Test the devices by pressing the test button or using a tester.

    - If a device malfunctions or clicks for no apparent reason, replace it with a compliant model.

## 6. Problems with lighting:

- **Causes:** Lighting systems may suffer from flickering, intermittence or failure to light due to faulty bulbs, damaged wiring or worn switches.

- **Repair:**

    - Check the voltage in the circuit and check each bulb.

    - Replace defective bulbs and check that the wiring is correct.

- If the problem persists, replace the switch or use a more stable power supply.

**Conclusions:** Addressing frequent failures with a methodical approach ensures quick and effective repairs. Correctly identifying the cause of failure is essential to avoid recurring problems and ensure plant safety.

In the next section, we will look at real cases of failures and the solutions adopted, providing a practical view of how to solve complex problems.

## 6.4 Analysis of real cases and solutions adopted

Analyses of real-world cases provide valuable perspective on the challenges electricians face on a daily basis and the solutions adopted to solve problems. These examples illustrate how a methodical approach and the right skills can prevent and solve complex failures.

**Case 1: Frequent tripping of residual-current circuit breakers** In one house, residual-current circuit breakers were tripping repeatedly, causing inconvenience to residents and interrupting power in several rooms.

- **Diagnosis:** The electrician made a visual inspection of the electrical panels and wiring, checking for faulty connections. He then isolated each circuit and measured the leakage current with a multimeter.
- **Solution:** A faulty appliance was found to be causing leakage and activating the differentials. The appliance was disconnected and replaced, resolving the problem.

**Case 2: Voltage drop in lighting systems** In a commercial building, the lighting system had a significant voltage drop, causing lights to flicker.

- **Diagnosis:** The electrician tested the voltage at several points in the system and identified a significant decrease from the nominal value. Inspection of the cables revealed that they were of insufficient cross section for the length of the route.
- **Solution:** The cables were replaced with larger cross sections, reducing voltage drop and ensuring stable power supply to the entire lighting system.

**Case 3: Short circuit in a multiple socket system** An office had experienced a sudden power failure due to a short circuit in a group of multiple sockets.

- **Diagnosis:** The technician isolated the circuit involved and tested the sockets one by one. A faulty connection was discovered in one of the outlets, which had caused a short circuit.
- **Solution:** The socket was replaced and the connections were checked and tightened. The circuit was tested again to make sure there were no further short circuits.

**Case 4: Cable overheating in** an industrial plant In an industrial plant, cable overheating had occurred, causing a partial blackout and damage to electrical panels.

- **Diagnosis:** The maintenance team used thermal imaging cameras to identify overheated cables, noting that the runs were overloaded and not well insulated.

- **Solution:** New cables with appropriate cross sections were installed, reducing the load on each path. Connections were improved with insulation materials and additional protection.

**Case 5: Interference in security systems** In a residential building, security systems had false alarms due to electromagnetic interference.

- **Diagnosis:** The electrician monitored the current variations in the safety system circuits and discovered that the cables were not shielded properly and were running too close to an electric motor.

- **Solution:** The cables were shielded with appropriate materials and relocated away from the source of interference, eliminating false alarms.

**Conclusions:** These real-life cases highlight the importance of accurate diagnosis and tailored solutions to prevent and solve problems in electrical systems. A methodical approach and knowledge of the correct techniques help electricians quickly restore systems to working order.

In the next section, we will look at a guide to small home repairs, providing practical tips for dealing with common problems users may encounter at home.

# Chapter 7: Common Repairs

## 7.1 Guide to small home repairs

Minor household repairs are essential to maintain the efficiency of electrical systems and prevent more serious failures. This guide provides practical tips for solving common problems you may encounter in your home.

**1. Before starting:**

- **Safety:** Before working on any electrical system, always disconnect power from the main switchboard. Check with a tester that the circuit is actually disconnected.

- **Tools needed:** Have tools such as insulated screwdrivers, wire strippers, voltage tester and multimeter available. Use insulated gloves to avoid accidental shocks.

- **Circuit knowledge:** Familiarize yourself with the house wiring diagram and identify which circuits power the various rooms.

**2. Socket repair:**

- **Loose sockets:** If a socket is loose, turn off the power, remove the front plate, and tighten the screws that secure the socket to the electrical box.

- **Sockets not working:** If an outlet is not working, check the associated residual current or circuit breaker. If it still does not work, turn off the power, remove the outlet, and check that the wires are securely connected.

- **Burned out sockets:** If a socket shows signs of burning, replace it immediately. Check that the current drawn by connected devices does not exceed the circuit capacity.

## 3. Switch repair:

- **Switches that do not work:** If a switch does not turn the light on or off, it may be worn or have loose connections. Disconnect the power supply, remove the switch, and check that the wires are well connected. If necessary, replace the switch.

- **Tripping circuit breakers:** If a circuit breaker keeps tripping, it could indicate an overload or short circuit. Identify devices connected to the circuit and distribute loads among multiple circuits.

## 4. Lighting:

- **Flickering bulbs:** If a bulb flickers, it may be defective or have a loose connection. Replace it and check that the bulb holder is in good condition.

- **Lamps that do not work:** If a lamp does not light, check the switch and voltage at the bulb holder. Make sure the wires are securely connected and replace the bulb if necessary.

## 5. Dispersal problems:

- **Residual current circuit breakers tripping:** If a residual current circuit breaker trips, it may indicate a leakage. Isolate connected circuits and devices one at a time to locate the source of the leakage. Replace defective wiring or devices.

6. **Preventive maintenance:**

- **Visual inspection:** Periodically check outlets, switches and electrical panels for signs of wear, damage or overheating.

- **Functional testing:** Regularly test residual current circuit breakers and protective devices. Check continuity and voltage of circuits with a multimeter.

**Conclusions:** Small household repairs require attention and a methodical approach to ensure safety. By addressing these problems in a timely manner, one can keep the system in optimal condition and avoid more serious failures.

In the next section, we will explore how to replace switches and outlets properly, ensuring that new installations are safe and compliant.

## 7.2 Replacement of switches and sockets

Replacing switches and receptacles safely and effectively is a critical activity for maintaining the integrity and efficiency of home electrical systems. Here is a practical guide to performing this task correctly.

1. **Preparation:**

- **Turn off the power supply:** Before starting any electrical work, be sure to turn off the power supply to the affected circuit from the main electrical panel. Use a tester to check that there is no voltage in the outlet or breaker.

- **Tools needed:** Obtain a set of insulated screwdrivers, wire strippers, multimeter and insulated gloves. If possible, also keep a crimping tool and cable ties.

2. **Removal of the switch or socket:**

    - **Removing the plate:** Remove the protective plate of the switch or socket using a screwdriver. Set the screws aside for reuse if in good condition.

    - **Removing the unit:** Pull the switch or outlet out of the electrical box and identify the connected wires. Note the order and color of the wires for reconnection.

- **Disconnecting the wires:** Use a screwdriver to loosen the terminal screws and disconnect the wires. If necessary, cut off the worn ends and strip the conductors again.

3. **Installation of the new switch or socket:**

    - **Cable connection:**

        - **Switch:** Connect the corresponding wires to the terminals of the switch following the original order. Make sure that the wires are tight in the terminals and that the insulation is not inside.

        - **Receptacle:** Connect the phase, neutral and ground wires to the corresponding terminals of the receptacle. Make sure each connection is securely attached.

    - **Placement in the electrical box: Once** connected, place the switch or socket inside the electrical box. Securely fasten the unit with the original or new screws.

    - **Mounting the plate:** Mount the protective plate on the switch or socket and secure it with screws.

4. **Testing and verification:**

    - **Restore power:** Turn on the power from the main electrical panel and test the operation of the switch or socket.

    - **Safety test:** Use a tester to check for voltage at terminals and make sure the circuit is working. Check for leakage or faulty connections.

5. **Future maintenance:**

    - **Periodic inspection:** Check switches and outlets periodically for signs of wear, overheating, or loose connections.

    - **Tightening:** Tighten terminal and plate screws if necessary. Immediately replace any unit that shows signs of damage.

**Conclusions:** Replacing switches and receptacles is a simple but crucial procedure for maintaining the integrity of the home electrical system. Making sure each connection is tight and insulated is essential to prevent failure and ensure safe use.

In the next section, we will look at how to approach small appliance repair, providing tips on how to diagnose and solve common problems.

## 7.3 Repair of small household appliances

Repairing small household appliances can seem like a challenge, but with the right tools and step-by-step guidance, you can fix common faults and extend the life of your devices. In this chapter, we will look at how to diagnose and repair common faults in three popular home appliances: the dishwasher, the television, and the refrigerator. These instructions provide a solid foundation for Section 7.4 below, which deals with work on lighting and lamps.

## Repairing a Dishwasher

**Tools needed:**

- Screwdrivers
- Multimeter
- Protective gloves

**Steps:**

1. **Diagnosis of the problem:** Check if the dishwasher does not start, does not drain water, or does not clean well. Observe if there are any error codes displayed on the panel.

2. **Disconnecting from power:** Always disconnect the device from power before proceeding with any repairs.

3. **Filter and pump check:** Remove and clean the filter. Check if the drain pump is clogged and clean it if necessary.

4. **Test the pump with a multimeter:** Use the multimeter to check the continuity of the pump. If it does not show continuity, it may need to be replaced.

5. **Reassembly and test:** Reassemble all parts and reconnect the unit to power. Run a short cycle to verify that the problem has been solved.

## Repairing a Television

**Tools needed:**

- Screwdrivers
- Multimeter
- Welder (if necessary)

**Steps:**

1. **Fault identification:** Determines whether the problem is related to power, sound, or image. Check for sound without image, which could indicate a backlight problem.

2. **Opening the TV:** Use screwdrivers to gently open the back of the TV.

3. **Power supply board check:** Use multimeter to test capacitors on the power supply board. Replace swollen or damaged capacitors.

4. **Check connections and cables:** Make sure all connections are firm and cables intact.

5. **Final test:** Reconnect the TV and verify that it is working properly.

**Repairing a Refrigerator**

**Tools needed:**

- Screwdrivers
- Cleaning brush
- Multimeter

**Steps:**

1. **Diagnosis of the problem:** Check if the refrigerator is not cooling properly or is making unusual noises.

2. **Cleaning the condenser coils:** Disconnect the refrigerator from power. Clean the condenser coils, which may be clogged with dust and dirt.

3. **Checking the fan motor:** Check the fan motor with a multimeter to make sure it is working properly. Replace it if it does not show continuity.

4. **Check thermostat:** Test the thermostat with a multimeter to make sure it is working properly. Adjust or replace it if necessary.

5. **Reassembly and verification:** After all repairs have been made, reassemble the refrigerator and reconnect it to test its operation.

These examples show how to address common problems in small appliances, offering practical solutions and boosting confidence in home maintenance skills. The knowledge gained here facilitates the transition to the next chapter on lighting and lamp repairs, maintaining a logical and consistent thread through the book.

## 7.4 Interventions on lighting and lamps

Work on lighting systems and lamps is common in home maintenance activities. Understanding how to address lighting issues can improve energy efficiency, increase safety, and optimize the overall lighting in the home.

1. **Diagnosis of common problems:**

- **Lamps not turning on:** Check the power switch and fuse panel or circuit breakers first. Check if the bulb is burned out by replacing it with a working one.

- **Lamps flickering or going out:** This can be caused by a loose connection, deterioration of the lamp holder, or problems with the power supply. Check connections and replace damaged components if necessary.

- **Circuit overload:** If the lamps go out when other devices are on, the circuit may be overloaded. Consider redistributing the load or installing an additional circuit.

## 2. Replacement of bulbs:

- **Selecting the appropriate bulb:** Choose bulbs that fit the fixture specifications, considering type (e.g., LED, incandescent, fluorescent), fitting size and maximum allowable wattage.

- **Bulb installation:** Always turn off the power before replacing a bulb. Make sure it is screwed on firmly but without excessive force to avoid damage to the bulb holder.

## 3. Repair of lamp holders and switches:

- **Damaged lampholders:** Damaged lampholders can cause connection problems. Disassemble the fixture, replace the lamp holder and reconnect the wires, making sure to connect the wires correctly according to the color code.

- **Faulty switches:** Switches can wear out over time. If a switch malfunctions, turn off the power, disassemble the switch and replace it with a new one that has the same specifications.

## 4. Installation of new lighting fixtures:

- **Planning:** Assess the installation site to ensure that there is adequate support for the new luminaire and that the location is optimal for the desired lighting.

- **Installation:** Follow the manufacturer's instructions for installation, connecting the wires (phase, neutral and ground) in the appropriate connectors. Securely fasten the unit to the ceiling or wall.

## 5. Safety measures and maintenance:

- **Regular inspection:** Make regular inspections to ensure that there are no exposed wires, loose connections, or signs of overheating in the light fixtures.

- **Cleaning of** luminaires: Keep luminaires clean of dust and debris to prevent overheating and maintain good lighting efficiency.

**Conclusions:** Proper maintenance and repair of the lighting system not only enhances the aesthetics of the home but also contributes to overall safety by reducing the risk of fire or electrical failure. Following proper procedures for installation and maintenance ensures that lighting is both efficient and safe.

In the next section, we will explore the precautions and practical safety tips that are essential for any work on household electrical systems.

# 7.5 Precautions and practical safety tips

Safety in the installation and maintenance of electrical systems is crucial. This chapter discusses how to properly handle cables and switches, emphasizing the importance of different cable colors and how they should be properly combined with switches and receptacles. This information sets the stage for the next section, 8.1, concerning basic circuit design.

## Types of Cables and Colors

Electrical cables are classified primarily according to their use and configuration, with different colors to facilitate identification and ensure secure connections:

1. **Single-phase cables:**
    - **Blue (Neutral)**: Blue cable is always used for neutral in all types of installations.
    - **Brown (Phase)**: This color indicates the phase cable, carrying current.

2. **Three-phase cables:**
    - **Black, Gray, and Brown**: These colors represent the different phases in a three-phase system and are used to distinguish the various phase lines.

3. **Green/Yellow (Earth):**
    - This cable is used for grounding and has the function of protecting against accidental electrical discharge. It is essential in all electrical installations to ensure safety.

## Combination with Switches and Sockets

To ensure proper installation:

- **Switches**: Always connect the phase (brown) wire to the switch. The switch will interrupt the flow of current into the phase, allowing safe control of the power supply.

- **Receptacles**: Wires should be connected respecting polarity; the neutral (blue) in the corresponding terminal usually marked with the letter N, the phase (brown) in the terminal marked with the letter L, and the ground wire (green/yellow) in the terminal with the ground symbol.

## Safe Use of Cables

- **Insulation and protection**: Ensure that all cables are properly insulated and undamaged before proceeding with any installation. Use protective conduits or raceways where cables are exposed to potential wear and tear.

- **Checking connections**: Check regularly that all connections are secure and tight to prevent malfunction or fire hazards.

- **Avoid overloading**: Do not overload electrical outlets with too many devices to avoid the risk of overheating and short circuits.

## Practical Example

Imagine installing a new socket in a room. Here are the steps:

- **Preparation**: Turn off the main power supply. Check with a voltage tester that there is no current.

- **Installation**: Connect the ground wire to the corresponding terminal, followed by the neutral (blue) and phase (brown).

- **Verification**: Once the socket is installed, turn the power back on and use a tester to verify proper installation.

These safety principles not only provide protection during installation and daily maintenance, but also provide the foundation for the next chapter on basic circuit installation design, ensuring that each step is performed with the utmost attention to safety.

# Chapter 8: Practical Projects

## 8.1 Basic circuit installation project

A basic circuit in electricity is the minimum structure required to power a device or equipment within a home. Typically, it includes a protective device (such as a fuse or circuit breaker), a phase conductor that supplies the current, a neutral that completes the circuit by returning to the power source, and often a ground wire for safety. These circuits are common in various household appliances, including refrigerators, lights, televisions, and small appliances such as blenders. The basic circuit is crucial because it provides the necessary path for the electrical energy to operate safely and efficiently.

**Practical Example: Installation of a Basic Circuit for a Blender**

A blender is a common household appliance that requires a safe and reliable electrical circuit. Here is how to go about installing a basic circuit for a kitchen blender.

**Tools needed:**

- Voltage tester
- Screwdriver
- Electric drill
- Cable clamps
- Clamp

**Materials:**

- Suitable electrical cable (three-wire cable with ground)
- Electrical outlet
- Built-in box
- Circuit breaker

**Steps:**

1. **Circuit planning:**
   - Determine the location of the nearest appropriate electrical outlet for the blender, ideally close to the point of use to avoid the use of extension cords.

2. **Installation of recessed box:**
   - Use the drill to install the recessed box in the designated spot. Make sure it is well secured and level.

3. **Laying cables:**
   - Stretch the cable from the electrical panel to the flush-mount box. Use cables that meet the current requirements of the blender and include a ground wire.
   - Make sure the cables are secured along the route and protected from mechanical damage.

4. **Socket connection:**
   - Connect the wires to the corresponding terminals on the new outlet (phase, neutral, and

ground). Use the voltage tester to make sure the outlet is connected correctly and safely.

5. **Installation of the circuit breaker:**
   - Install a circuit breaker appropriate for the intended load in the electrical panel. This will protect the circuit from overloads and short circuits.

6. **Circuit testing:**
   - Turn the power back on from the main panel and use the voltage tester to check that the outlet and circuit are working properly. Check that the blender is working properly with no signs of overload or malfunction.

This example illustrates a methodical and safe approach to installing a basic circuit, which not only serves a blender but is applicable to many other small appliances. Understanding and applying these principles is critical to proceeding to Section 8.2 below, which deals with the design and implementation of an alarm system, extending the skills learned in handling safe and reliable circuits.

## 8.2 Design and implementation of an alarm system

Home security is a priority for many homeowners, and installing an effective alarm system is an essential way to protect the home from intruders and other threats. A well-designed alarm system not only provides security but also adds value to the property. In this segment, we will explore how to design and activate a home alarm system and also suggest three types of alarms that offer excellent value for money.

**Steps for installing a home alarm system**

1. **Needs assessment:**
    - Determine the most vulnerable areas of the house, such as front doors, ground floor windows, and other easily accessible access points.

2. **Choice of alarm system:**
    - Consider systems that offer a good balance between cost and functionality. Here are three recommendations:
        - **Wireless alarm system:** Easy to install and configure, ideal for those seeking a simple and less invasive solution.
        - **Wired alarm system:** More reliable in terms of connection, but requires professional installation.

- **Smart alarm system:** Integrates home automation features, controllable via smartphones or other smart devices.

3. **Installation of sensors and cameras:**
   - Place motion sensors in strategic locations such as near entrances and in high-traffic areas.
   - Install security cameras to monitor and record activities at critical locations around and inside the house.

4. **Control panel configuration:**
   - Install the control panel in an area that is easily accessible but not too exposed. Make sure it is connected to a power source and, if possible, to a telephone line or Internet connection.

5. **Activation and testing of the system:**
   - Activate the system following the manufacturer's instructions. Test each component to make sure it is working properly, checking that all sensors are responding and that the cameras are clearly capturing images.

6. **Training and information:**
   - Familiarize yourself with the operation of the system. Learn how to disable the alarm in case of a false alarm and how to react if the alarm goes off.

## Practical Example: Activating a Wireless Alarm System

1. **Planning:**

- Decide on the number and location of sensors based on the floor plan of the house.

2. **Installation:**
   - Attach the sensors to doors and windows using adhesives or screws provided by the manufacturer.
   - Place the control panel in a central location, such as the entrance hallway.

3. **Configuration and connection:**
   - Follow the instructions to connect the control panel to your home Wi-Fi and synchronize it with all sensors and cameras.

4. **Testing:**
   - Activate the system and try opening a door or window to test the sensors. Also check the response of the cameras.

This approach provides reliable protection for the home and is easily integrated with additional technological improvements, such as garage or office automation, which will be explored in the next section 8.3 of the book.

## 8.3 Garage or small office automation

Automating a garage or small office not only increases convenience and efficiency, but can also improve safety and energy efficiency. This chapter guides you through the process of implementing automation solutions, from automated garage doors to smart lighting and security systems, providing a practical example that serves as the basis for Section 8.4 below, devoted to electrical retrofit of a room.

## Most Popular Types of Automation

1. **Garage door automation**: Includes systems that allow automatic opening and closing via sensors or remote controls.

2. **Security systems**: Include security cameras, motion sensors, and alarm systems that can be monitored and controlled remotely.

3. **Smart lighting**: Uses bulbs and lighting systems that can be programmed or controlled via apps, motion sensors, or voice assistants.

4. **Smart climate control**: Smart thermostats that optimize energy consumption by adjusting the temperature according to the time of day or the presence of people.

## Practical Example: Garage Automation

**Tools needed:**

- Garage door automation kit
- Drill
- Screwdriver
- Smartphone or tablet for app configuration

**Steps:**

1. **Choice of Automation Kit:**
    - Select a garage door automation kit that is compatible with your existing model and

supported by an app with good reviews for ease of use and reliability.

2. **Installation of the Device:**

    o Follow the instructions provided with the kit to mount the automated device on the existing garage door. This usually includes attaching a drive unit to the garage ceiling and connecting it to the door opener.

    o Install any safety or limit switch sensors that let the system know when the door is fully open or closed.

3. **Electrical Configuration and Connection:**

    o Make sure all electrical connections are safe and up to code. Use cables of appropriate cross section to connect the motor unit to the power supply.

    o Connects the system to the home Wi-Fi infrastructure to enable remote control.

4. **App configuration:**

    o Download and install the mobile application related to your automation system.

    o Follow the instructions to synchronize the app with your new garage door automation system. Configure user preferences, such as automatic opening times and motion notifications.

5. **System Testing:**
    - Test the system to make sure the garage door opens and closes properly via app commands.
    - Check the functionality of sensors and security settings to ensure that everything is working as expected.

6. **Maintenance:**
    - Schedule periodic checks to verify the mechanical and electrical integrity of the system and to ensure that the app is up to date.

This example of garage automation not only improves functionality but also safety, making everyday life more efficient. This approach can be replicated or adapted for small offices, with the addition of similar automations such as lighting and air conditioning, facilitating the transition to Section 8.4, which further explores the electrical retrofit of indoor environments.

## 8.4 Electrical retrofit of a room

An electrical retrofit involves upgrading the electrical systems of a room or an entire building to improve their efficiency, safety, and functionality, adapting them to modern consumer and technological needs. This process can range from the simple replacement of obsolete devices, such as switches and outlets, to a complete renovation of the wiring system. A well-executed retrofit not only increases safety and energy efficiency, but also increases property value. Below, we detail how to perform an electrical retrofit in a room.

## Materials and Tools Needed:

- Electrical cables: select copper or aluminum cables of appropriate cross-section, conforming to current regulations.
- Modern electrical switches and outlets: choose models with additional features such as surge protectors or USB ports.
- Upgraded switchboard: if necessary, replace the old switchboard with one that supports more circuits and circuit breakers.
- Voltage tester and multimeter: to check for current and test various electrical parameters.
- Screwdriver, pliers, electric drill, hole saw: for physical installation of elements.
- Insulated gloves and safety glasses: to protect the operator during work.

## Steps for Electric Retrofit:

1. **Assessment and Planning:**
   - Inspects the existing system and identifies items that need to be replaced or upgraded.
   - Design the new electrical layout based on current needs, considering the addition of modern devices and an increased need for power points.
2. **Power disconnection:**

- Be sure to disconnect the power supply to avoid any risk during work. Verify the absence of voltage using a tester.

3. **Removal of obsolete facilities:**
    - Carefully remove old wiring, switches, outlets, and the electrical panel if necessary. Document the existing configuration to help with the reinstallation process.

4. **Installation of new wiring:**
    - Install the new cables following the designed layout. Be sure to use conduit or protective sheathing where necessary to protect cables from mechanical damage.

5. **Installation of sockets and switches:**
    - Mount the new outlets and switches in the designated locations. Use the drill and hole saw to create openings in the walls if necessary. Connect the cables, observing the polarities indicated.

6. **Switchboard configuration:**
    - Install a new electrical panel if required, connecting and configuring circuit breakers for each circuit. Clearly label each circuit to facilitate future work.

7. **System testing and verification:**
    - Once the installation is complete, reconnect the power supply and test each component with the

multimeter. Verify that all elements are functioning properly and that the system meets safety regulations.

8. **Final review and cleaning:**

    o Perform a final check of all installations to make sure there are no exposed cables or loose connections. Clean the work area by removing debris and residual materials.

This electrical retrofit process not only modernizes the system to meet today's technological needs, but also prepares the facility for future upgrades, ensuring that the electrical system is safe, reliable, and up to code. This solid, up-to-date foundation is essential for the transition to the next point in the book, which deals with the automation of environments such as garages or offices.

# Chapter 9: Innovations and the Future of Electricity

## 9.1 Home Automation and Smart Home

The world of electricity is undergoing a radical transformation thanks to technological innovations that affect both the way electricians work and the way we live in our homes. This chapter explores some of the emerging technologies in the home sector and provides step-by-step instructions on how to implement them effectively.

Home automation solutions have revolutionized home comfort and energy management, making homes smarter and more interconnected. Devices such as smart thermostats, app-managed security systems, and remotely controllable lighting are now commonplace in many modern homes.

**Installation of an Intelligent Thermostat:**

**Tools needed:**

- Screwdrivers
- Multimeter
- Smartphone or tablet for configuration

**Steps:**

1. **Turn off the power:** Make sure the heating/cooling system is turned off from the main switchboard to avoid accidents.

2. **Remove the old thermostat:** Disconnect the wires from the existing thermostat, making sure to label them to facilitate the installation of the new device.

3. **Connect the new thermostat:** Attach the wires to the new thermostat following the manufacturer's specific instructions. Use multimeter to check for proper power supply.

4. **Configure the device:** Turn on the thermostat and use the associated app to configure it according to your temperature and program preferences.

5. **Test the system:** Checks that the thermostat correctly adjusts the temperature and interacts with the heating/cooling system.

## 9.2 Renewable Energy

Installing renewable energy systems such as solar panels has become a popular choice to reduce energy costs and increase independence from traditional power grids.

**Installation of Solar Panels:**

**Tools needed:**

- Solar panel kit
- Installation tools (drill, wrenches, etc.)
- Multimeter for testing connections

**Steps:**

1. **Site assessment:** Determines the best location for solar panels, usually the roof, where they receive maximum solar exposure.

2. **Installation of supports:** Attach the panel supports to the roof, making sure they are well anchored and aligned.

3. **Mounting the panels:** Install the panels on the supports, connecting them together according to the manufacturer's instructions.

4. **Inverter connection:** Connects the panels to the inverter, which converts solar energy into usable electricity in the home.

5. **Home grid connection:** Integrate the solar system with the home's existing electrical system and install any monitoring systems.

## 9.3 Electric Vehicle Charging Stations

The adoption of electric vehicles has led to the need for home charging stations that are convenient for owners of such vehicles.

**Installation of a Charging Station:**

**Tools needed:**

- Charging station kit
- Electrical installation tools
- Voltage tester

**Steps:**

1. **Choose location:** Identify an accessible spot in the garage or in a protected outdoor area near where you park your vehicle.

2. **Electrical installation:** Connect the charging station to the main power supply, following the technical specifications to ensure adequate power and safety.

3. **Setup:** Configure the charging station following the manufacturer's instructions, including configuration of any monitoring apps.

4. **System test:** Verifies that the station is working properly by connecting the vehicle and monitoring the charging process.

## 9.4 Intelligent Lighting

LED technology and smart lighting systems offer ways to reduce energy consumption and improve the home atmosphere.

**Installation of Intelligent Lighting:**

**Tools needed:**

- Intelligent LED lamps
- Screwdriver
- Smartphone or tablet

**Steps:**

1. **Removal of existing lighting:** Replace old bulbs with smart LED lamps.
2. **Configuration via App:** Connect the lamps to an app on your smartphone to control lighting intensity, color, and scheduling.
3. **Test the system:** Verify that all lamps work as programmed and are easily controlled via the app.

## 9.5 Home Automation

Home automation integrates various technologies, from security to energy efficiency, into a remotely controllable system.

**Installation of a Home Automation System:**

**Tools needed:**

- Home automation kit
- Installation tools
- Smartphone or tablet

**Steps:**

1. **Device selection:** Choose devices that are compatible with your home and can be integrated into the automation system.
2. **Physical installation:** Install motion sensors, cameras, and other control devices at strategic locations in the home.

3. **System configuration:** Connect all devices to a central unit that manages them. Configure settings via an app on your smartphone.

4. **System test:** Verify that all automation system components are communicating properly with each other and with your control device.

These examples illustrate how technological innovations can be implemented to improve daily life by making homes safer, more efficient, and more comfortable.

# Frequently Asked Questions in the Field of Electrician with Practical Examples

### 1) How to handle electromagnetic interference in the home?

**Frequently Asked Question:** What are strategies to minimize electromagnetic interference in home environments, especially with the increasing presence of electronic devices?

**Practical Solution:**

- **Tools needed:** EMC (Electromagnetic Compatibility) filters, shielded cables.

- **Steps:**

    1. **Identify sources of interference:** Understand what devices (such as microwaves, Wi-Fi routers, and other electronics) can cause interference.

    2. **Use of EMC filters:** Install EMC filters on the most problematic devices to mitigate the emission of interference.

    3. **Optimized cabling:** Uses shielded cables to reduce interference pickup, especially in areas of high device density.

    4. **Intelligent arrangement of devices:** Place sensitive equipment away from those that generate the most interference.

5. **Verification and testing:** Checks periodically the effectiveness of the adopted solutions, adapting strategies as the home technology environment evolves.

## 2) How do you integrate automation with home firefighting?

**Frequently Asked Question:** How can I improve my home fire protection system by integrating automation technologies?

**Practical Solution:**

- **Tools needed:** smart smoke detectors, home automation hub.

- **Steps:**
  1. **Choose smart smoke detectors:** Install smoke detectors that can connect to Wi-Fi and communicate with a central app or hub.
  2. **Integrate with automation systems:** Connect detectors to an automation system to allow immediate notifications and automatic actions such as activating emergency lights or closing gas valves.
  3. **Simulations and tests:** Run regular tests to make sure the system responds appropriately in an emergency.
  4. **Ongoing Maintenance:** Check and replace detector batteries as recommended by the manufacturer to ensure continued functionality.

### 3) Installation of energy monitoring systems

**Frequently Asked Question:** How can I monitor and manage energy consumption in my home?

**Practical Solution:**

- **Tools needed:** Energy monitor, energy management app.
- **Steps:**
    1. **Select an energy monitor:** Choose a device that can measure energy consumption in real time and is compatible with smartphones or computers.
    2. **Monitor installation:** Connect the monitor to the main electrical panel.
    3. **App setup:** Set up the monitoring app to display consumption data, set alarms for over-consumption, and receive recommendations to reduce energy costs.
    4. **Data analysis:** Use collected data to identify and reduce unnecessary consumption, optimizing equipment use and improving energy efficiency.

### 4) Implementation of electrical solutions for gardens and outdoor areas

**Frequently Asked Question:** What electrical solutions can I implement in my garden to improve its aesthetics and functionality?

**Practical Solution:**

- **Tools needed:** Outdoor lights, automated watering systems, outdoor cables.
- **Steps:**
    1. **Plan lighting:** Determine the areas of the garden that would benefit from lighting for aesthetic or safety reasons.
    2. **Install outdoor lights:** Choose lights suitable for outdoor use and place them along paths, near structures or as accent lighting.
    3. **Automate irrigation:** Install a programmable irrigation system that activates according to weather conditions or at set times.
    4. **Maintenance and upgrades:** Maintain systems regularly to ensure that they operate effectively and safely, making improvements or upgrades when necessary.

# **Conclusion**

As we come to the end of this journey through the pages of "The Electrician's Handbook: The Ideal Companion for Every Technician," it is time to reflect on the importance of what we have learned and the transformation this knowledge can bring to the daily practice of every electrical professional and enthusiast. This handbook was not just a set of instructions; it was a true exploration into the heart of electricity, a detailed guide to safely and competently navigating a field as vast as it is vital.

The value of this book lies in its ability to transform theory and technical information into practical, applicable tools that can be immediately used in both professional and domestic settings. From the fundamentals of electricity, through innovative home automation technologies, to the design and implementation of complex electrical systems, each chapter has been structured to provide not only the theory but also detailed, practical examples that illustrate each step needed to implement effective and safe interventions.

We hope that, on page after page, you have found the answers you were looking for and that you now feel better prepared to face any electrical challenge with new confidence and increased competence. The world of electricity is constantly changing, and with this handbook in your possession, you are ready to adapt and thrive, taking advantage of the latest technologies and best practices in the industry.

We now invite you to take a moment to reflect on how this book may have influenced the way you work or your approach to electrical problems. If you found the content and advice offered helpful, please consider sharing your experience by leaving a review. Your impressions can guide other readers in their choice and help the author improve further editions of this valuable handbook.

Thank you for accompanying us on this exploration of the world of electricity. Continue to use this book as a trusted companion in your electrical adventures, and never stop expanding your knowledge and honing your skills. The future of electricity is bright, and with the right tools, yours can be too.

*If you think you liked this book and it helped you, I only ask you to take a few seconds to leave a short review on Amazon!*

*Thank you,*

*Filippo Ferrari*

www.ingramcontent.com/pod-product-compliance
Lightning Source LLC
Chambersburg PA
CBHW072017230526
45479CB00008B/100